AA

100
Marvels
of the
Modern World

AA Publishing

Written by Alison Ahearn, Andrew Forbes, Fay Sweet and Hamish Scott

Produced by AA Publishing

Maps and text © Automobile Association Developments Limited
Relief maps created from originals supplied by Getty Images/ The Studio Dog and Mountain
High Maps®, Copyright © 1993 Digital Wisdom

Published by AA Publishing (a trading name of Automobile Association Developments
Limited, whose registered office is Fanum House, Basing View, Basingstoke, Hampshire
RG21 4EA. Registered number 1878835).

ISBN-10: 0-7495-4801-0
ISBN-13: 978-0-7495-4801-4

A02320

The AA's website address is **www.theAA.com/bookshop**

A CIP catalogue record for this book is available from the British Library.

The contents of this book are believed correct at the time of printing. Nevertheless, the
publishers cannot be held responsible for any errors, omissions or for changes in the details
given in this book or for the consequences of any reliance on the information provided by
the same. This does not affect your statutory rights.

Design layouts by Liz Baldin at BOOKWORK Creative Associates, Hampshire, England
Origination by Keene Group, Andover
Printed and bound in Dubai by Oriental Press

AA

100
Marvels
of the
Modern World

Contents

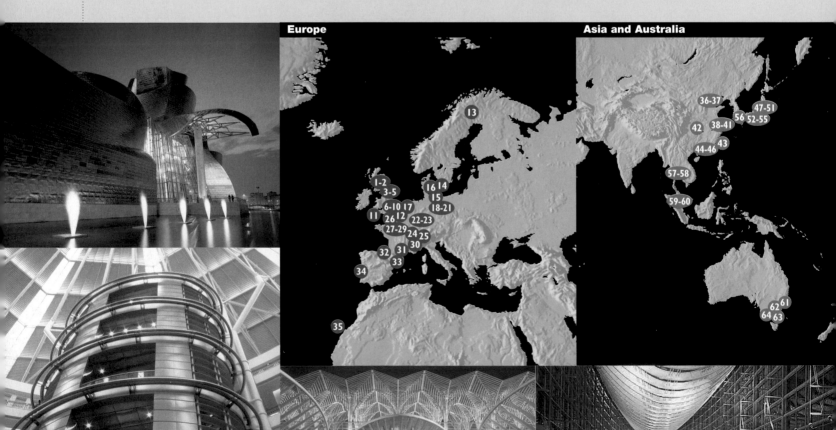

Europe

Asia and Australia

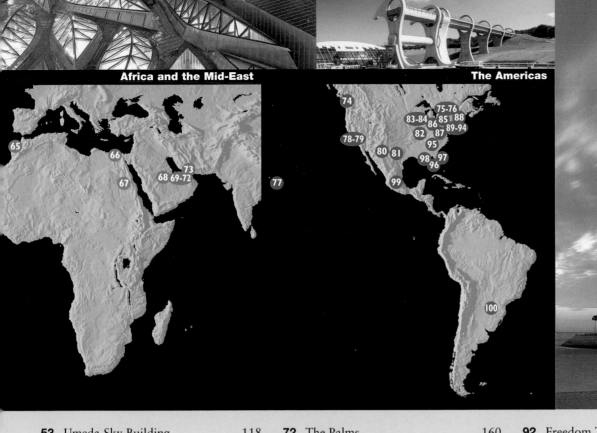

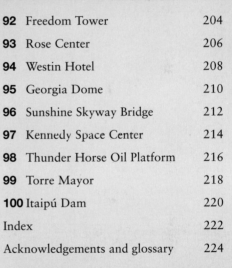

Africa and the Mid-East

The Americas

Europe

I n previous centuries, European monarchs commissioned the construction of great cathedrals and sumptuous palaces, knowing that the work would take so many decades (and even centuries), that they would never see the masterpiece completed. Now it is unusual for any construction project to take more than a decade. Such is our demand for the new and innovative, that multi-million dollar budgets and construction teams working round the clock are the norm. The difference is, that while utility and beauty of form are still paramount, the twenty-first century technologies available to the architects and engineers of today are cutting-edge.

Marvels of engineering in this chapter range from the awe-inspiringly massive construction project, the Channel Tunnel, which is the physical embodiment of the special relationship (or *entente cordiale*) between Britain and France. Another way to cross a divide in spectacular style, is by driving across France's Tarn Valley via the Millau Viaduct. Only inaugurated in 2004, it is a masterpiece of form and function and its breathtaking beauty is matched only by the fact that it was completed on time and within budget. Denmark's Nysted Windfarm, like almost all modern energy-generating projects, has faced (largely unjustified) criticism of its impact on the environment, yet it is difficult not to admire the grace and serenity of its form. The architectural antithesis to Nysted, is Berlin's Jewish Museum, where memories of the Holocaust are deliberately kept alive by grating, uncomfortable design elements that seek to upset and unnerve its visitors.

Europe is known for the fact that it contains such a great diversity of architectural styles in such a small geographic area, and the marvels in the following pages only add to this tradition. From luxury hotels to ice hotels, giant wheels to giant sculptures, tall towers to tidal barriers, Europe still has vision in the fields of architecture and engineering.

Each of the 112 steel nodes in the debating chamber roof weighs half a ton and is individual in its design.

The building is clad in 7,200sq yards (6,000sq m) of Kemnay granite from Aberdeenshire, in Scotland.

Each office has an area by the window, described by the architect as "a contemplative space".

Scottish Parliament

From the moment the Scottish Parliament was conceived it caused controversy. The project's architectural design and cost have been subjects of furious debate, and its 4-acre (2ha) site at Holyrood, on the edge of Edinburgh's medieval Old Town, is certainly a challenging location for a public building. Squeezed between a royal palace and the city's densely built-up lanes, the Parliament can be seen only from close quarters or from the massive cliffs of Salisbury Crags and Arthur's Seat, hills which tower above the city.

Architecturally, there are no concessions either to tradition or to modern principles of rational design. When the Catalan architect Enric Miralles (1955–2000) first sketched out possible ideas, he began with drawings of flowers, leaves, and hills rather than with conventional plans. Much of the building as it stands is a structural interpretation of these themes.

Organic Principles

The Parliament lies in city streets, yet in place of formal grounds, branching earthworks cloaked in moorland grass form a seamless link with the natural landscape of the crags. Towers resembling upturned boats, skylights that are veined like leaves, curving walls, and turf roofs reveal the structure as a strange organic form that, as critics have suggested, owes more to artistic genius than common sense.

Spiraling Costs

The project proved notorious for going wildly over budget, from an initial estimate of US$72 million to a final cost of US$776 million. A commission headed in 2003 established a variety of causes. The first figure was certainly too low for a building of such original design, but aside from this the required floor space increased by a third, and more than 15,000 changes were made to the plans. Further problems were caused by the sudden death of the architect Enric Miralles in July 2000.

Bespoke Design

The building's unusual design and quality of finish becomes apparent in the public foyer, where concrete vaulting has been polished to the smoothness of marble. On the upper level, the debating chamber forms an amphitheater bathed in natural light that pours through half-screened windows and enormous skylights. The chamber roof is an astonishing construction. Its 33-yard (30m) span is supported on enormous beams of laminated oak braced with slender bars of stainless steel and jointed with handmade steel nodes.

Behind the debating chamber building, four towers house committee rooms and other parliamentary facilities. Elsewhere in the labyrinthine complex, the building housing offices for Members of the Scottish Parliament (MSPs) is of an entirely different design. Here, each office, resembling a monastic cell, has been constructed as an individual vaulted unit. The windows are particularly unusual, as each was handmade to an elaborate abstract design to provide a curiously shaped seating alcove that has acquired the nickname of a "think-pod." One last surprise, providing a contrast to the Parliament's modernity, is the handsome Queensberry House which dates from the 17th century. It has finally been restored to its former glory and is now used as accommodation for MSPs.

A Home at Last

Scotland's independent parliament was dissolved in 1707, following political union with England. In a referendum held in 1997, however, 74.3 percent of Scottish people voted in favor of devolution, retaining Scottish MPs in Westminster, London, while matters of purely Scottish interest were settled by a separate parliament in Edinburgh. At first the project proceeded at a hectic rate—a site was chosen from a choice of four locations and a design was selected from 70 submissions by the end of 1998. The first elections were held in 1999, but due to construction problems the parliament remained in temporary accommodation until the Holyrood buildings were finally ready in October 2004.

Falkirk Wheel

As the world's first rotating boatlift, the Falkirk Wheel, near Rough in Falkirk, Scotland, was innovative in terms of engineering and design. As a machine it functions with superb efficiency, raising and lowering boats between two canals with a minimal expenditure of energy, but it is also a structure of dramatic beauty that can be appreciated as a massive, mobile sculpture.

In 1998, when British Waterways launched the Millennium Link, a project to reopen a coast-to-coast canal link across Scotland, the greatest engineering challenge to the scheme lay in forging a connection between the Forth and Clyde Canal and the Union Canal at Falkirk. The 115ft (35m) difference in height between these waterways had previously been negotiated through a flight of 11 locks, but this was demolished in 1933 as the canals fell out of use. Rather than rebuild the cumbersome old structure, the project's backers commissioned a creative team of architects and engineers to come up with a more imaginative, contemporary solution that would modernize the image of the waterway and establish its success. Initially, various options were considered, from a giant Ferris wheel to tilting tanks, but after three weeks of discussion the principle of a boatlift was agreed. Construction started in the summer of 2000 and the wheel was opened in May 2002.

HOW IT WORKS

The Falkirk Wheel's efficient operation relies on the Displacement Principle discovered by the Greek inventor Archimedes in the 4th century BC. As each boat enters one of the steel gondolas it displaces its own volume of water, so that both gondolas retain a consistent mass of 430 tons (300 tonnes). Power derives from 10 hydraulic pumps driven by a small electric motor connected to a drive gear, 26ft (8m) in diameter, which is fixed to the pier closest to the aqueduct. Two smaller gears connect this with a further pair of cogs that embrace the gondolas within the structure's arms, ensuring absolute stability during rotation.

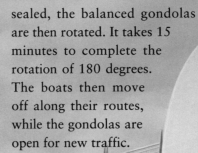

Art and Engineering

The Falkirk Wheel is designed not only to be practical, but also to provide both users and onlookers with a sense of drama and excitement. Boats on the upper-level Union Canal emerge from a newly cut 158-yard (145m) tunnel onto a 327-yard (300m) aqueduct that marches out from the hillside, supported by five sets of concrete piers. When viewed from a boat, this aqueduct appears to end in mid-air. It is only from below that the true situation is apparent. The final section of the aqueduct is in fact a steel gondola, large enough to hold a pair of 65ft (20m) boats, and sealed by lock-gates at both ends. Meanwhile, on the lower level, a second gondola admits boats from a mooring basin on the Forth and Union Canal. With their lock-gates sealed, the balanced gondolas are then rotated. It takes 15 minutes to complete the rotation of 180 degrees. The boats then move off along their routes, while the gondolas are open for new traffic.

THE MILLENNIUM LINK PROJECT

The opening of the Falkirk Wheel marked the completion of the Millennium Link, a project that has restored the historic waterways between the Firth of Forth and the River Clyde. The canals, which once carried coal, goods, and passengers between Edinburgh and Glasgow, fell into decline with the growth of railways. By the mid-1960s the route was a disused and derelict eyesore. Restoration involved cutting new lengths of canal and raising a section of the busy M8 freeway. The revived waterway now provides a focus for both leisure activities and commercial development.

The Arts Council of England and the European Regional Development Fund financed the sculpture.

The sculpture is made of steel to which copper has been added, causing it to oxidize to a rich red-brown color.

The wings and body were brought from Hartlepool and assembled using a 560-ton (500-tonne) crane.

Angel of the North

Gazing far into the distance across the A1 trunk road, the monumental figure of an iron angel with outstretched wings greets visitors to Tyneside in the north of England. For many of the 90,000 people who stream past each day in cars and trains, Britain's largest sculpture has become a familiar feature of the landscape since it was erected in 1998. But for those who have never caught a glimpse of it before, its presence on the skyline comes as an astonishing surprise. With feet firmly planted on the turf, the figure's massive form stands 65ft (20m) tall. Its wingspan, at 175ft (54m), is only 21ft (6m) less than that of a Jumbo jet. Although recognizably an angel, the semi-abstract figure is very much a product of its time and place. Built above a disused coal mine and constructed of raw steel, the angel represents both Tyneside's industrial identity, and the region's cultural renaissance in the 21st century.

On one level, the Angel of the North can be appreciated purely as a work of art, devoid of any practical associations. But the sculpture can also be admired from a more down-to-earth perspective as a magnificent idea, dreamt up by an artist, which has been realized using modern engineering skills and local industrial traditions.

Larger than Life

The Angel started life in Anthony Gormley's London studio as a small-scale model, which was then recorded as a 3-D computer image. Working from this "virtual" angel, civil engineers Ove Arup & Partners then worked out how the full-scale sculpture might be built to last at least a century and stand up to winds of above 100mph (160kph). When the plans were complete, the wings and body were constructed at a steel-works in Hartlepool while the site above Gateshead was carefully prepared.

Although the Angel seems to stand delicately balanced on the grass, 52 steel bolts, each 10ft (3m) long, secure its feet to a buried concrete plinth and foundation piles extending 65ft (20m) down to bedrock. The sculpture is entirely built of steel, with an outer skeleton of ribs, welded to a "skin" 1/4in (0.5cm) thick. Within the hollow body there's a spine-like core, more ribs and five massive transverse plates. Two of these plates support the wings, weighing 56 tons (50 tonnes) apiece, while the whole sculpture weighs in at 224 tons (200 tonnes). Colossal, controversial, and brilliantly designed, the Angel has become an iconic symbol of the north of England.

GODDESS OF THE NORTH

In 2005 plans were announced for the creation of an even larger artwork by the A1 road, 15 miles (24km) to the north of the Angel. Designed by the American writer and landscape artist Charles Jencks, the "Goddess of the North" will be constructed from the spoil of an opencast mine. The recumbent female figure, representing the Celtic water-goddess Coventina, will have hips and breasts up to 100ft (31m) high, forming a landscape modeled on the human form.

The figure will continue to grow as coal is extracted from the mine and, once the seam has been exhausted, the site will be opened as a public park.

ANTHONY GORMLEY

Born in 1950, sculptor Anthony Gormley achieved international recognition for works on a far smaller scale than the Angel of the North. In 1991 he worked with a family of Mexican brick-makers, using several tons of clay to produce 35,000 individual figures 3in–10in (8cm–26cm) tall for a tableau entitled *Field*. In 1994 he won the Turner Prize for a British version of this work, which was made with the help of 100 children and their families. In an on-going project, he continues to produce new "Fields," collaborating with communities from around the world.

the **facts**

The hydraulic system is powered by eight electric motors generating 589 horsepower.

When the bridge is raised, the suspension cables are kept taught by 14-ton (16-tonne) castings to either side.

The bridge is self-cleaning, with litter rolling into chutes at either end each time the structure tilts.

Millennium Bridge

GATESHEAD QUAYS

The Millennium Bridge is part of a regeneration project worth US$908 million that has transformed formerly run-down quays along the River Tyne into a lively hub of art and entertainment. Architecturally, the most spectacular development is Norman Foster's Sage Centre, which includes a 1,650-seat concert hall, a secondary auditorium, rehearsal rooms and a music school within a complex "shrink-wrapped" beneath a stainless-steel roof. Near by, the Baltic Mill is a 1950s grain store that has been cleverly converted into an exhibition space and arts center, while a number of apartments and hotels are also under construction.

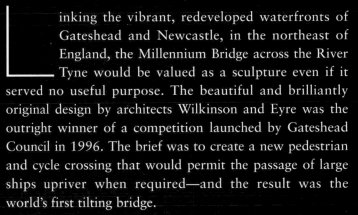

L inking the vibrant, redeveloped waterfronts of Gateshead and Newcastle, in the northeast of England, the Millennium Bridge across the River Tyne would be valued as a sculpture even if it served no useful purpose. The beautiful and brilliantly original design by architects Wilkinson and Eyre was the outright winner of a competition launched by Gateshead Council in 1996. The brief was to create a new pedestrian and cycle crossing that would permit the passage of large ships upriver when required—and the result was the world's first tilting bridge.

The bridge consists of two steel arcs across the river. One, under normal use, is raised 165ft (50m) above the water, while the other lies horizontal, carrying the cycleway and footpath in a sweeping curve from bank to bank. In this position there is sufficient clearance for small craft to pass beneath, but when a larger ship approaches the whole structure is rotated on its axis through 40 degrees, with the carriage deck slowly rising as the upper arch descends. The process, which takes four-and-a-half minutes, is like the gradual blinking of an eye and once it is complete, with both arcs raised and counterbalancing each other, there is clearance of 82ft (25m) for the shipping lane below.

Drama and Performance

The bridge was a major engineering project, costing US$40 million and taking more than two years to complete. Having been fabricated at a works in Bolton, Lancashire, the steel structure was delivered as a kit of parts for assembly at a yard at Wallsend, on the mouth of the Tyne. The complete bridge was then carried 6 miles (10km) upriver on Europe's largest floating crane, the *Asian Hercules*. At times the enormous structure, 138 yards (126m) wide and weighing 952 tons (850 tonnes), had to be rotated to fit the river's width. Remarkably, when the bridge was at last lifted into its prepared position, it fitted with just one tenth of an inch (2mm) to spare.

A year later, in September 2001, the bridge gave its first public performance, opening before an audience of 35,000 onlookers. Since then it has tilted around 200 times a year, with each occasion causing crowds to stop and stare. But even when motionless, the bridge still attracts attention. Designed both as a river crossing and as a theatrical experience, it has redefined the image of Tyneside.

TYNE BRIDGES

Close to the new river crossing, three much older bridges span the Tyne, all of which were engineering marvels of their time. The High Bridge, built between 1846 and 1849 by British engineer Robert Stephenson, was the world's first double-decked bridge, carrying a road with a railway above. Next to it, the Swing Bridge of 1873 rotates around its central pier to allow ships to pass. Soaring above both of these is the great arch of the Tyne Bridge, dating from 1928. Measuring 435 yards (398m) from side to side, it was the largest single-span bridge in the world until the opening of Sydney Harbour Bridge in 1932.

With more than 65,000 individual plants, the garden has one of the largest collections of European plants in Britain.

The water for the Grand Cascade is recycled, with 250,000 gallons (1,135 liters) constantly stored in underground tanks.

In 2003 the gardens attracted 530,000 visitors.

Alnwick Gardens

The magnificent new water-gardens of Alnwick Castle, near Hexham in Northumberland, are as grandiose as any to be found in the grounds of Britain's historic stately homes. In the 18th and 19th centuries it was not uncommon for ducal families to spend fortunes on elaborate fountains and parterres, but owing to the cost of maintenance few examples have survived, and modern gardens tend to be both more informal and constructed on a far more modest scale.

Alnwick's gardens are not, however, some extravagantly self-indulgent throwback to the past. Not only are they modern in design, but, in the spirit of a democratic age, they are intended to be enjoyed by as many as possible.

The project was conceived in 1996, when the Duchess of Northumberland decided to create a contemporary water-garden of international significance on 12 acres (5ha) of derelict walled gardens and abandoned nurseries within the grounds of Alnwick Castle. A year later the Belgian landscape designer Jacques Wirtz was commissioned to produce a plan, and construction work started in March 2000. In September 2002 the first phase of the gardens was ready to be opened to the public and was opened by HRH the Prince of Wales. With ownership now transferred to a charitable trust, the gardens continue to evolve and are due to be completed by 2008. In the meantime, work on the gardens continues to progress.

GRAND CASCADE

Alnwick Garden's Grand Cascade is one of the largest and most sophisticated water features of its kind to be built in recent years. At peak flow, 7,260 gallons (33,400 liters) of water per minute pour down its series of 21 weirs, to be drawn through four large, bell-shaped openings into powerful computerized pumps that control the fountain display. The three largest jets reach a height of 20ft (6m), 40 smaller jets reach 13ft (4m), while 80 side-jets throw parabolas of water from the sides of the cascade. There are four different combinations of display, with their sequence changing every half hour.

Grand Designs

Centered on a Grand Cascade of tumbling weirs and jetting fountains, the gardens celebrate the beauty and the power of water. Aside from the cascade's spectacular displays, intricate water-sculptures in the Serpent Garden demonstrate intriguing illusions and effects, from vortices and wave patterns to hydrostatic tension. Rills of water also bubble through the Ornamental Garden, radiating from a central pool through a geometric scheme of paths and hedge-lined "rooms." The layout is designed to provide a series of distinct experiences for the visitor, with a labyrinthine bamboo maze, a Poison Garden of poisonous, healing and narcotic plants, and a Rose Garden of 3,000 bushes. There is also a 667-sq yard (558sq m) complex of treehouses, linked by aerial walkways.

Among other features planned for the coming years is a Garden of the Senses, in which children will wear blindfolds to explore the natural world using only touch and smell, with water providing interesting sounds. Combining plantsmanship with education, art and entertainment, the gardens appeal to modern tastes in representing a new form of gardening design.

DIGGING DEEP

Before work started on the gardens, a team of 30 archeologists spent three weeks surveying the site. They found traces of six former gardens dating back to the 18th century, when the first Duke of Northumberland employed the English landscape gardener Lancelot "Capability" Brown (1716–83) to redesign the park. The gardens were at their grandest in the 1890s, with acres of flowers, five grape houses, five pine houses, and a conservatory. The formal gardens were abandoned after World War II and the site was used as a tree nursery. The concept for the gardens as they are now, came from the present Duchess of Northumberland.

The hollow gates, which are filled with water, are faced with steel up to 2in (5cm) thick.

Power comes from three independent sources, with three generators and a manual system as back-up.

Two subways cross the river through the concrete sills, providing access to the piers and lifting gear.

Thames Barrier

The line of huge steel-plated structures that strides across the River Thames at Woolwich could, with just a few hours' warning, save London from disaster. Housing the Thames Barrier's hydraulic lifting gear, the gleaming, sail-like forms are the most visible parts of a sophisticated machine designed to protect the city from the potential devastation of a tidal surge. If such a flood occurred, there could be an incalculable loss of life. The danger was brought home in 1953, when a storm-driven surge killed thousands in the Netherlands and more than 300 on England's east coast.

For two decades there was little in the way of practical response. It was only after London's docks moved out to Tilbury in the early 1970s that the construction of a barrier was seen to be viable. Even then, the river had to remain navigable for the large ships that occasionally sailed up the river beyond Tower Bridge. The solution finally adopted was to build the world's largest moveable barrier. It took eight years to construct, but since its completion in 1982 it has proved effective in defending London without disrupting the river traffic which still makes its way up and down the River Thames every day.

THE FUTURE

Despite the presence of the barrier, the long-term risk of floods remains and may be increasing due to climate change. Sea levels are predicted to rise by up to 3ft (1m) by 2100, and there appears to be an increase in both the severity and frequency of storms. The barrier was closed in response to a flood threat on just four occasions during its first five years of operation. Since then, closures have increased to a record 13 times in 2001 and 11 in 2003. Official estimates predict a further rise to 75 a year by 2050 unless further measures are put in place.

FREAK CONDITIONS

Tidal surges are quite commonplace events that cause harm only when exceptional circumstances coincide. They occur when a band of low pressure crosses the Atlantic, raising a low "hump" of water just a few inches high but several hundred miles across and traveling at 50mph (80kph). As this surge approaches the Norwegian coast, a strong north wind may drive it down into the shallow waters of the North Sea, where its intensity will increase enormously, raising normal water levels by up to 10ft (3m). If the surge also coincides with an extreme high tide, its effect on estuaries and coastlines can be devastating.

Holding Back the Tide

The barrier consists of 10 gateways that stretch 569 yards (522m) from bank to bank between nine concrete piers. The two outer gates to either side are simple sluices, while the remaining ones are designed for navigation. Two are 34 yards (31m) wide and suitable for small craft, while the four central channels are each 67 yards (61m) in width, the same as Tower Bridge. The gates themselves are normally invisible, recessed into concrete sills on the river-bed, but their vital statistics are in keeping with the colossal forces that they have to resist. Weighing up to 1,680 tons (1,500 tonnes) and designed to withstand more than 10,000 tons (9,000 tonnes) of pressure, each is bolted to a counterbalanced steel disc that is in turn connected to the arm of a massive rocker-beam that provides motive power.

The order for closure is usually given four hours ahead of a predicted surge tide. Starting with the outer gates, the arms rotate the discs, which slowly raise the gates through 90 degrees until there is a wall of steel across the river. The closure is deliberately gradual to avoid a reflective wave rebounding back upriver, although the barrier could potentially be sealed in just nine minutes.

The wheel and capsules, weighing 1,120 tons (1,000 tonnes), turn on a cantilevered steel spindle 27 yards (25m) long.

Each capsule has its own motorized stability system, keeping it horizontal at all times.

The "flight" takes 30 minutes.

London Eye

Towering 443ft (135m) above the South Bank of the River Thames in London, this enormous observation wheel has become one of the city's most familiar landmarks. Remarkably, the project was conceived without official backing or even the support of a developer. The husband-and-wife team of architects responsible for its design, David Marks and Julia Barfield, drew up initial plans around their kitchen table and campaigned tirelessly to see their vision realized before striking up a partnership with the airline British Airways.

The battle to obtain planning permission took two years and, in the light of the site's sensitive location opposite the Palace of Westminster, the matter was even discussed in Parliament. But once the go-ahead was given, the London Eye was completed in just 16 months, despite the fact that almost every component and technique involved had to be invented from scratch. The giant wheel proved hugely popular from the moment it opened on the eve of the millennium. More than four million people take a "flight" each year, making it the city's third most popular attraction after the British Museum and the Tate Gallery.

Revolutionary Design

Although based on the concept of the Ferris wheel, the London Eye is a new and far more sophisticated form of observation wheel. Instead of hanging from the rim as they revolve, the 32 air-conditioned capsules stand clear of the perimeter so that views are not interrupted by the structure as they rise. With the wheel moving at a constant speed of

LIFT OFF

During the construction process individual sections of the Eye were carried up the Thames on barges and assembled in a flat position on platforms of steel piles. Once it was complete, the wheel was then hoisted upright, using four jacks pulling 144 steel strands to raise it at a rate of two degrees an hour. Before it was edged into its permanent position, the wheel remained at 65 degrees for a week while final adjustments were made. At the time, it was believed to be the largest object ever raised from a horizontal to a vertical position.

MILLENNIUM DOME

The London Eye's success contrasts with Britain's official monument to the millennium, the Millennium Dome. Designed by Richard Rogers, the dome was built to house an exhibition celebrating modern culture and technology. However, as costs spiraled the project faced a hostile press and increasing public disapproval. The dome attracted 6.5 million visitors during the year that it was open (compared to the 12 million originally predicted), but since its closure in December 2000 the enormous structure has remained virtually unused. It is, however, likely to become a venue for gymnastic events at London's 2012 Olympics.

10in (26cm) per second, passengers step into the capsules from a ground-level platform while those who have already made the trip alight. Up to 800 people can be carried at a time, but without the need to stop and start, the finely balanced mechanism consumes only 0.5 watts (500Kw) of electricity, costing just a few cents per passenger. Since its opening, the success of the London Eye has inspired rival projects in Las Vegas and Singapore that may in time displace its world pre-eminence in terms of size, but it will remain one of London's best-loved examples of architectural exuberance. Although planning consent was originally granted for only five years, rumors of its possible closure were greeted with dismay in 2005, provoking intervention from London's mayor.

The steel in the arch weighs 1,929 tons (1,750 tonnes), as much as 275 London double-decker buses.

Construction of the stadium has included 275 miles (444km) of mains electrical cabling.

The stadium houses four of the largest restaurants in London and 98 kitchens.

Wembley Stadium

WEMBLEY ARCH

As well as being an iconic landmark, the distinctive arch straddling the stadium, set at 68 degrees from horizontal, is an important structural feature. It fully supports the north side of the roof together with 60 percent of the south side, and is integral to the sliding mechanism of the roof. Four times higher than the original stadium's twin towers, it spans 344 yards (315m)—the combined length of three soccer pitches—and has a diameter of 23ft (7m), wide enough for a Channel Tunnel train to run through.

S eating an impressive 90,000 spectators, packed with state-of-the-art facilities, and with a soaring 436ft (133m) triumphal arch, which can be seen for miles across London, the new Wembley Stadium was the largest soccer stadium in the world when it opened in 2006. It was designed by the World Stadium Team, a joint venture between the world-renowned architect firm Foster and Partners and the global stadium specialist HOK Sport, and built by Multiplex Constructions, who were responsible for Stadium Australia in the 2000 Olympic Games.

A National Sports Ground

During the 1880s there were soccer pitches at Wembley Park along with running tracks, cricket pitches, and gardens. After World War II the Government planned a British Empire exhibition on the site with a national sports ground as its centerpiece. The Empire Stadium, as it was originally known, was built in just 300 days and opened in 1924. This building, with its distinctive twin towers, eventually changed its name to Wembley Stadium and became the home of English soccer. It was the main venue of the 1948 Olympic Games, and in 1966 hosted the final of the World Cup won by England. After struggling for years to meet modern needs, it finally closed in 2000 to make way for the new structure.

Sight and Sound

The massive building is circular in shape, wrapped in aluminum and glass, and partially roofed with two crescent-shaped membranes to protect spectators from bad weather. Weighing 8,000 tons (7,000 tonnes), the roof is partly retractable, and is left open between events so that the famous Wembley turf can be kept in prime condition by being exposed to natural sunlight and ventilation.

Replacing the four stands of the former stadium, steeply raked seating is arranged in a single bowl-shaped auditorium. Spectators can also watch close-up action on a pair of massive screens, each the size of 600 television sets. One of the distinctive features of the former stadium was its acoustics, which produced the famous Wembley Roar from cheering crowds. Before it was demolished in 2000, acoustic specialists made recordings in the old stadium and worked closely with the architectural team to ensure the familiar sound would remain.

While designed primarily for soccer, rugby league and music events, the stadium is highly versatile and can be reconfigured to host world-class athletics events. On these occasions a prefabricated platform is imported to sit over the pitch and lower levels of seating, to make a field large enough for track and field events. The stadium will be the venue for the soccer finals in the London 2012 Olympics.

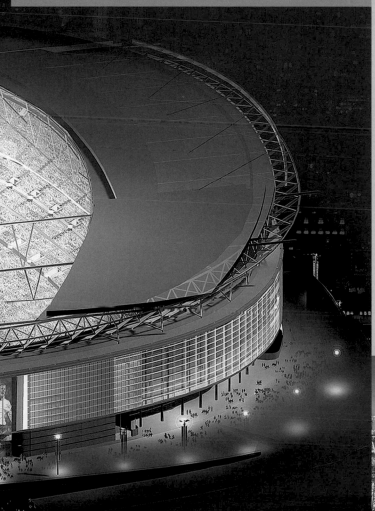

the facts

- The building was officially opened in March 2004 and it can, in theory, accommodate about 4,000 people.
- The building contains nearly 22 miles (35km) of structural steel.
- The glass cladding consists of 5,500 panels, covering the equivalent of five soccer pitches.

9   30 St Mary Axe   **London,** England

30 St Mary Axe

When 30 St Mary Axe first emerged above London's city skyline in 2003, it was unlike any tall building that Londoners had ever previously seen. It almost instantly became known as "the Gherkin," and, although the likeness to a pickled cucumber might seem a touch far-fetched, the nickname does acknowledge not only the tower's unique, organic shape, but also its proud claim to be "green." The building's well-known architect, Lord Foster, has described it as a "pine cone."

Quite aside from aesthetic factors, the shape allows the 40-story office building to occupy a minimum of ground space. From a diameter of just 55 yards (50m) at base level, it expands to a bulbous waist of 61 yards (56m) on the 16th floor before tapering towards its highest point. In fact, the building's height of 590ft (180m) is only 7ft (2m) greater than its maximum circumference. This means that most of the site is left free as public space, a valuable commodity in London's densely built financial heart. The shape also offers less resistance to the wind, reducing the down-draughts and eddies that swirl around tall structures.

High-rise London

Following the completion of the 600ft (183m) Nat West Tower (now Tower 42) in 1980, it was more than 20 years before any new high-rise buildings were built in the City of London. Lord Foster submitted plans for a 90-story Millennium Tower on the site now occupied by 30 St Mary Axe, but they were refused. Had it been built, it would have been Europe's tallest building. Following the sale of the site to the international Re insurance company, Swiss Re, Foster's smaller, more acceptable alternative "Gherkin" design was accepted.

Natural Ventilation

The building consumes up to 30 percent less energy than an equivalent-sized office block of conventional design. Much of this saving is achieved through the tower's "biomorphic" form, which reduces reliance on artificial lighting and air conditioning. The atria that spiral up around the core open up the structure to sunlight and fresh air, which is drawn by natural convection through the diagonally rising levels. Windows can also be opened to allow natural ventilation of the building.

Spirals and Diamonds

The building is constructed in an unusual and very visually exciting way. Although elevators and services are located in a central core, the tower's strength comes from its external skeleton of steel tubes. This forms a latticework of giant diamond shapes which encompasses the circular facade. Within these diamonds, the horizontal steel rings that support individual floors are clearly visible through triangular, glazed panels.

Nor is this the limit of the tower's geometrical complexity. One diamond in every three is glazed with darker glass, producing great diagonals that spiral upwards from street level, giving the impression that the building has been twisted on its axis. In fact, the darker bands enclose multistory atria that rotate around the tower's core, bringing light, air, and open views to balconies on every floor. Despite the building's shape, the only piece of curved glass is the lens at the top of the building. Enjoying 360-degree panoramic views across the city and towering above St Paul's Cathedral, this spectacular glazed area houses London's highest bar, which is set aside for the exclusive use of tenants and their guests.

The tower will be able to accommodate around 7,000 people at any one time.

39 high-speed elevators, plus 12 escalators for the lower levels, will serve the tower.

The 34th floor will contain London's highest swimming pool, 400ft (122m) above street level.

London Bridge Tower

By the time of the 2012 Olympics, London's skyline will have been transformed by a number of eye-catching new buildings. Of these, the most spectacular by far will be the London Bridge Tower. Planned for completion by 2010, the 1,016ft (310m) "shard of glass" will be Europe's tallest building, and it looks likely to become London's foremost landmark. The building's height will not be its only claim to fame, for the structure will be architecturally innovative.

Designed by the Italian architect Renzo Piano, the soaring, spire-like tower will be encased in "extra-white" glass that will take on the color of ambient light so that, depending on conditions, its form may become a shaft of light blazing in the sun or, at other times, seem to vanish in thin air. When viewed from close quarters, steeply raked angles pointing to infinity will add to the impression of enormous height, while geometric cutaways through the multilayered glass facades will provide a complexity of detail rarely found in modern buildings of such a scale. At the top of the tower, above the 70th floor, an open latticework of sail-like devices will function as "wind radiators," directing breezes down between the layers of glass to provide a system of natural air conditioning.

Vertical Living
Unlike any other London high-rise building, the tower will be a mixed development that functions 24 hours a day. Its site, by London Bridge Station, is one of the capital's busiest transport hubs and the lobby will be entered from a new public square, sheltered under canopies. The first three levels will consist of stores, with 26 floors of offices above. A further 18 floors will contain a luxury hotel and health spa, while upper levels will house apartments, some of which may occupy whole floors. What is more, there will be public spaces interspersed throughout the structure. Midway up, between the office levels and hotel, there will be a triple-floor plaza with restaurants, bars, an exhibition

gallery, and perhaps even a chapel. But the tower's greatest draw will be the top-floor gallery, with views stretching for 30 miles (50km) across the city and beyond. Rising from a station that serves thousands of commuters every day, and replacing horizontal streets with high-speed elevators, the building will symbolically point London in a new direction—a city reaching for the skies.

MIXED REACTIONS

The tower sparked a furious debate when plans were first announced. At a public inquiry in 2003 that lasted for seven months, English Heritage claimed that it would "pierce the heart of London like a glass spike" and usurp the traditional pre-eminence of St Paul's Cathedral. Renzo Piano, on the other hand, described his creation as "a sharp and light presence in the London skyline." Supporters of the scheme, including London's mayor Ken Livingstone, claimed that it would improve the city's image as a dynamic, modern capital. The tower finally received a go-ahead from John Prescott, Britain's deputy prime minister, in November 2003.

RENZO PIANO

As one of Europe's most respected architects, Renzo Piano is admired for both the originality and sensitivity of his designs. Born in Genoa, Italy in 1937, he studied architecture at the Polytechnic of Milan before forming a partnership with the British architect Richard Rogers in 1971. During the seven years they worked together, the pair achieved international fame for the Pompidou Center in Paris, one of the first examples of high-tech design.

More recently, Piano has experimented with contrasting styles on projects that range from Kansai Airport in Japan to the wonderfully romantic Tjibaou Cultural Center on the exotic South Pacific island of New Caledonia.

The framework of the biomes consists of 625 hexagons, 16 pentagons, and 190 triangles.

952,000 tons (85,000 tonnes) of bark and compost were blended with china-clay waste to create the soil.

All imported plants go into quarantine to ensure no diseases or pests are introduced.

the facts

Eden Project

Since opening in 2001, the complex of giant greenhouses near St Austell in Cornwall, England, has seized the imagination of the world, attracting well over a million visitors a year. Constructed in a former clay-pit, the transparent, interlocking domes appear to bubble from the ground like science-fiction life-forms, clinging weightlessly to the quarry's cliff-face. The largest biome, made up of four domes up to 180ft (55m) high and covering 4 acres (2ha), is large enough to contain the Tower of London. Inside, the artificial climate replicates the humid tropics with a lush rainforest of trees, vines, waterfalls, and steaming pools. A second biome, covering 2 acres (1ha), has warm, temperate conditions, with plants from southern Europe, South Africa, and California. There are plans for a third, even larger, biome with an environment of arid desert.

Gardens with a Difference

The project was the brainchild of Tim Smit, a former music promoter who moved to Cornwall in 1987. He was already noted for his restoration of the Victorian gardens of Heligan nearby. From the start, Eden was conceived to be different from conventional botanical collections. Enclosed within a space-age structure, examples of the world's natural environments are presented as if gathered in an ark, safe from exploitation and destruction. Eden is designed to inspire a sense of wonder, and to stimulate awareness of our duty to protect the natural world.

Eden's architecture contributes hugely to the project's magical appeal. Designed by Sir Nicholas Grimshaw, the British architect responsible for London's Eurostar terminal at Waterloo, the biomes are lightweight, self-supporting structures that, despite their colossal scale, sit gently on the natural contours of the site. When seen from outside they are stupendous works of architecture and miracles of engineering science, yet from within they are virtually invisible. All that can be seen of their structure is a spider's web of delicate, geometric forms outlined against the sky.

THREATENED SPECIES

The Eden Project contains more than 135,000 individual plants and trees, representing almost 6,000 species. These include rare specimens from the Seychelles and South America that are under threat through changes to their habitat, and a pharmaceutical collection demonstrates the importance of biodiversity to modern medicine. Many of the trees in the humid tropics biome grow at an extraordinary rate in the jungle-like conditions. The tallest, a kapok tree (*Ceiba pentandra*) which was planted as a 39ft (12m) specimen in 2000, had already reached a height of 101ft (31m) by 2005.

A Natural Environment

In place of walls and level floors there are cliffs and rough, uneven slopes, which are overgrown with vegetation. The biomes artificially maintain tropical temperatures of up to 95°F (35°C), but the main source of heat for the plants is the sun, with the cliff-face acting as a storage radiator. The plants themselves also control the climate by producing more moisture as temperatures in the biomes rise.

THE STRUCTURE

The design of Eden's biomes is based on the geodesic dome, a revolutionary building system invented by the visionary architect Richard Buckminster Fuller (1895–1983). An outer skeleton of tubular steel hexagons and pentagons, each up to 12 yards (11m) wide, is bolted to an inner frame of triangles and hexagons to create a self-supporting structure that combines lightness with extraordinary strength.

Each hexagon contains a 7ft (2m)-deep "pillow" made from three layers of ETFE, a co-polymer discovered in the 1980s that is highly transparent, self-cleaning and impervious to ultra-violet light. Most importantly of all, it is just 1 percent the weight of glass.

10 million cubic yards (8 million cu m) of spoil were excavated, which is more than three times the volume of the Great Pyramid of Giza in Egypt.

20 million travelers use the tunnel every year.

Two crossover caverns are provided for emergencies, allowing trains to switch tunnels if either line is blocked.

CHANNEL TUNNEL RAIL LINK

Since the tunnel's opening, Britain's rail network has struggled to cope with the Eurostar trains, which are capable of a top speed of 180mph (288kph). The first section of a high-speed line was completed in 2003 and by 2007 68 miles (109km) of new track will cut the journey-time between London and Paris to two hours. The first mainline railway to be built in Britain since 1899 will have its own ultra-modern terminus linked to St Pancras Station. Aside from a new tunnel under the Thames Estuary, 11 miles (18km) of the route through London will be under the ground, avoiding further disruption.

Channel Tunnel

On 1 December, 1990, Graham Fagg, an English workman on the Channel Tunnel, reached through an opening he had cut into the chalk and shook hands with Frenchman Philippe Cozette. It was a defining moment in the history of both nations, but particularly significant for Britain, ending 8,000 years of isolation from the European continent. The tunnel was equally stupendous as a feat of engineering. It took seven years to build, involved 15,000 workers and

was said to be the largest construction project ever undertaken without the assistance of slave labor.

Stretching 32 miles (51km) from Folkestone, on England's south coast, to Sangatte, on France's north coast, two 9-yard (8m) wide rail tunnels run in parallel. A third, somewhat smaller, service tunnel runs in between. Although surveys had established the existence of a band of firm chalk marl that was relatively impervious to water, when work started on the service tunnel in December 1987

DREAM COME TRUE

The earliest known scheme for a Channel Tunnel dates from 1802. Gas-lit and ventilated by enormous chimneys, the proposal was little better than a fantasy, and English fears of an invasion ensured that it was not pursued. Later in the century a number of more practical projects were proposed and in 1881 tunneling actually began, but as late as 1930, Britain's military forces successfully opposed each and every scheme.

Hope was renewed in 1974 when tunneling began following Britain's entry to the European Union, but within 12 months an economic crisis had forced the project to be cancelled. Eurotunnel, a private-sector company, finally raised the required finance for the scheme in 1987.

many details of the geology en route still remained unknown. Also, despite advanced navigation systems using satellites and lasers, there were fears that the French and British sections of the tunnel might not coincide. In fact they were off target by just 14in (35cm) horizontally and 2.5in (6cm) vertically.

March of the Machines

Within the constricted confines of the tunnels, huge Tunnel Boring Machines (TBMs) advanced ahead of the construction crews, churning through the chalk with tungsten teeth at a rate of up to 5.5 yards (5m) an hour. Behind the TBM's rotating head, hydraulic rams kept the bit pressed into the chalk-face, while "gripper rams" retained the tunnel walls as segments of the concrete lining were installed. Each worm-like, cylindrical machine could extend up to 22 yards (200m) as it advanced, sheltering workers within its armored sections and discharging spoil onto a conveyor belt behind. In one record-breaking week, a British TBM advanced 464 yards (426m), although, allowing for delays, the average rate of progress was around one third this distance.

Despite setbacks that added hugely to the project's costs, the Channel Tunnel was officially opened on May 6, 1994, when trains carrying Queen Elizabeth II and President Mitterrand met midway between France and Britain.

Close to the Icehotel there's an Ice Chapel, which is often used for weddings and christenings.

Jukkasjärvi, the village where the Icehotel is built, means "meeting place" in the local Sami language.

The ice is crystal-clear due to the purity of the River Torne and the fast-flowing current.

Icehotel

Without doubt one of the world's most unusual and captivating places to stay, the Icehotel, 125 miles (200km) inside the Arctic Circle, attracts thousands of overnight guests and day visitors during its brief winter season. Each year the hotel is built from scratch from blocks of ice and snow. Guests sit on ice chairs, sleep on ice beds, and enjoy a nightcap in the Absolut Ice Bar where they drink vodka on the rocks from glasses made of ice. Even the hotel's movie theater screen is made of ice. Daytime activities include snowmobiling and ice sculpting.

Inspiration for the hotel began in 1989, when a group of Japanese ice artists visited the Swedish village of Jukkasjärvi and created an exhibition in ice. The following year a French artist arrived and staged an exhibition in an ice gallery named the Arctic Hall. When a group of visitors stayed overnight at the gallery, slept warmly and comfortably on reindeer pelts, then enthused about their experience, the idea began to take shape.

Building Blocks

Preparations for the rebuilding of the hotel begin each spring, when blocks of ice from the river are cut and stored in a vast freezer, ready for the next season's hotel. Tractors, ice saws, and ice tools, developed in Jukkasjärvi, are used. Construction begins in earnest in October, when a vaulted steel frame is put in place for the main part of the building, and snow canons and heavy moving equipment mold the snow over the frame. After two days the steel vaulted sections are removed and placed in a new location. Next, ice columns are put in place to give extra support to the now freestanding snow vaults.

By early December the main structure nears completion and refinement of the interior begins. With a constant indoor temperature of about 23°F (–5°C), the indoor working conditions are relatively comfortable compared to the outside temperature, which can drop below -40°F (-40°C). Late into the evening, sculptors saw and chisel the ice blocks to create windows, doors, and columns along with desks, beds, chairs, tables, lamps, and sculptures. Finally, handpicked national and international guest artists are invited to design the interiors of some of the rooms.

Winter visitors begin to descend on the hotel from around mid-December, and reservations are accepted until around the end of April. When the roof starts dripping the season is officially over and the Icehotel returns to where it originally came from—the river. In the winter of 2000 the Icehotel company opened another venture on the other side of the Atlantic Ocean, Icehotel Québec in Canada. The ready availability of ice and snow made the idea possible.

SLEEPING TIGHT

Every guest room has ice walls, ice ceiling, ice floor and an ice bed topped with a mattress and reindeer skins. Guests climb into their super-insulated thermal sleeping bags, which can protect against cold of -13°F (-25°C). There is also a sauna on hand to warm guests up in the morning after taking a drink of hot lingonberry juice.

SAMI CULTURE

The Icehotel is the perfect place to explore the culture of the indigenous Sami people, also known as Lapps, who have inhabited this harsh wilderness for thousands of years. Originally semi-nomadic and living by hunting and fishing, the Sami now live more settled lives in their native Lapland area, which spans Sweden, Norway, Finland, and Russia. One of their main sources of income is from herding reindeer.

The Sami culture is rich in folklore and handicrafts. Of particular note is the stunning national dress, decorated with vividly colored ribbons woven in abstract patterns.

Two controllers monitor conditions, and whenever wind speeds exceed 60mph (97kph) the bridge is closed.

The Link has 1,000 fire detectors, 223 closed-circuit television cameras and 178 emergency telephones.

In June, 2000, just before the official opening, 80,000 runners took part in a half-marathon across the Link.

the facts

Oresund Link

Just 9 miles (15km) of sea separates Denmark's capital, Copenhagen, from the coast of Sweden. There was a time, back in the 17th and 18th centuries, when the waters of the Oresund formed a barrier between the two nations, which were in an almost constant state of war. More recently, however, both countries have forged closer ties and in July 2000 they were physically united with the opening of the Oresund Link, one of Europe's most stupendous civil engineering projects.

Danish and Swedish governments first agreed to link the two countries in 1991, and a consortium of companies from Denmark, Sweden, France, and Britain designed and built the Link, which consists of three distinct elements. First, traveling to the east from Denmark, a 2-mile (3km) tunnel of concrete sections resting on the sea-bed takes the road and railway out to Pepperholm, a manmade island 3 miles (4km) long. This island is itself a remarkable construction, formed from sand and stone dredged from the sea-bed. However, the most visually exciting element is the High Bridge that spans the final 5 miles (8km) to the Swedish shore.

Harmonious Connections

Carrying both road and railway, the bridge is double-decked, with motor traffic on the upper level. On its completion in 1999 the central section was the longest cable-stayed structure of this nature in the world, with a clear span of 534 yards (490m) across the Oresund's main navigation channel. The four monumental pylons that support this

span are wonders both of engineering and of art. Constructed of concrete, they taper up to 667ft (204m) above the sea. In the misty conditions that are common on the Baltic they often disappear into the clouds, leaving the great cables mysteriously suspended like the strings of an enormous harp. The musical symbolism of the design is appropriate, for the bridge has helped to harmonize the communities on either side. In 2004 nearly 12,000 vehicles a day used the bridge, along with 17,000 rail passengers, benefiting the economies of both Copenhagen and Malmo.

PIECE BY PIECE

Construction of the Link required prefabrication on a massive scale. The tunnel was cast in 20 sections, each one 575ft (176m) long and weighing 62,000 tons (55,000 tonnes), that were cast in a yard on Copenhagen Harbor, towed into position by a fleet of tugs and then lowered into a trench excavated from the sea-bed. The 51 piers of the approaches to the High Bridge were brought from Malmo Harbor, although the deck girders came by ocean-going barge from Cadiz in Spain. The pylons for the High Bridge were cast on site, and its four deck sections were raised into position using temporary towers and a floating heavy-duty crane.

MARINE CONSERVATION

Construction of the Link was permitted only on the understanding that the ecology of the Baltic should not be disturbed. The bridge is aligned to cause a minimum reduction to the flow of tides and spillage from dredging was limited to just 5 percent of excavated sea-bed material. The manmade island, Pepperholm, has been used as the test-bed for a valuable ecological experiment. Left deliberately bare of vegetation, it now supports more than 300 plant species, along with amphibians, butterflies, and spiders. It is also home to Sweden's largest breeding colony of the rare little tern.

77 miles (123km) of undersea cable link the windfarm to Denmark's power grid, supplying 145,000 homes.

The turbines start automatically when winds reach 9mph (15kph) and cut out at speeds above 56mph (90kph).

An equivalent fossil-fuel power station would produce more than 552,000 tons (500,000 tonnes) of CO$_2$ a year.

the facts

Nysted Windfarm

On days when sea-mist does not obscure the view, you can see giant windmills on the horizon from the seaside village of Nysted in southern Denmark. By day the windmills appear as a forest of slender, matchstick towers that signal to the distant shore with endlessly rotating arms. By night they form a galaxy of twinkling lights. Although more than 6 miles (10km) from the shore, the huge structures cannot be ignored as they form one of the largest and most modern offshore windfarms in the world.

But if the turbines look impressive from the land, it is only from a boat that their true scale is apparent. Seventy-two enormous towers rise out the sea, covering an area of 9sq miles (24sq km). Each tower is 230ft (70m) tall, with 130ft (40m) blades taking its full height to 360ft (110m), the size of a 35-story

LOCAL CONCERNS

The inhabitants of Nysted initially opposed the windfarm, fearing its effect on tourism and sailing. Visitor numbers have not, however, been much affected and some 5,000 yachtsmen still call in at the pretty village every summer. On a coast noted for its nature reserves and wildlife, there were also strong concerns expressed by conservationists, but again the actual effect appears to have been minimal. Radar monitoring has established that seabirds navigate with ease between the turbine rows or simply fly around the site, while observation of seal colonies shows no decline in seal numbers.

skyscraper. With a total output of 165 megawatts, Nysted set a new record for offshore power production when its generators came on line in 2003. Despite having eight fewer turbines than Denmark's other windfarm at Horns Rev off the Jutland coast, the smaller site, with larger 2.3-megawatt turbines, generates over 10 percent more electricity.

Harnesssing the Wind

With winds at sea typically 50 percent stronger than on land, there's a good case to be made for offshore windfarms in terms of practical efficiency, but ideal sites are hard to find. At Nysted, the Rodsand sandbanks meant the towers could be built in shallow waters, 20 to 30ft (6–9m) deep, several miles from land in an area where winds are steady but gales relatively rare. Nonetheless, more than 130,000 cubic yards (100,000cu m) of the sea-bed had to be dredged up so the foundations could be laid and construction begun. Remarkably, the erection of the turbine towers, which were transported to the site on a giant floating crane, was completed in just 79 days. The whole project was finished by December 2003, a month ahead of schedule, and well within the budget of US$262 million.

Although the windfarm is unmanned, requiring only periodic maintenance, sophisticated electronic systems constantly monitor its operation. If the wind direction alters, the turbine heads automatically rotate to be face-on to the breeze, while in gustier conditions the rotors can be slowed both by feathering the blades and by applying brakes to keep a safety limit of 17 revolutions per minute.

A DANISH FIRST

For more than a century Denmark has been in the forefront of wind-power technology. In 1891 a school teacher named Poul la Cour (1846–1908) built the world's first wind turbine at the Askov High School, and by 1918 the country had 120 small wind-powered generating stations serving rural areas. Interest declined after the end of World War I, but revived in World War II when supplies of coal and oil were disrupted. By the time of the oil crisis in the 1970s, Denmark was a world-leader in terms of wind-power expertise, and the country currently has more than 4,000 turbines on land and sea, producing 20 percent of its power needs.

The Art Axis is 164 yards (150m) long and is the largest gallery in Denmark.

The land on which Arken stands was reclaimed from the sea during the late 20th century.

HM Queen Margrethe of Denmark opened the museum on 15 March, 1996.

Arken Museum

With its graceful "prow" pointing skyward, the Arken Museum of Modern Art at Ishøj, 12 miles (20km) south of Copenhagen in Denmark, is reminiscent of a moored ship. The flat coastal landscape was integral to architect Søren Robert Lund's design, and he sought to create a building that interacted with the surrounding coast landascape with its lakes, beaches, and harbors.

The prow marks the museum's entrance. Across the threshold, visitors enter the main foyer with its domed skylight and a huge block of Norwegian granite. Finished with matt, rough, and smooth-polished surfaces, this stands as a monument to the Ice Age. Also here is a large mosaic designed to resemble a compass, which shows the plan of the museum.

Inside the building the maritime theme is continued with metal walkways and balustrading, walls leaning at unpredictable angles as if you are at sea, and small-scale industrial details such as bulkhead-style lighting, a couple of porthole windows, and chunky nuts and bolts. Internal spaces range from intimate rooms to wide-open galleries and include exhibition space, a 280-seat theater, a 156-seat movie theater, a restaurant, and a gift store.

THE ART COLLECTION

Arken's collection includes Danish, Nordic and international art, with special emphasis on contemporary art from 1990 onwards. The works follow two themes—one is concerned with the portrayal of modern man, while the other focuses on new forms, materials and media, such as video and installation. Along with paintings, sculpture, and graphic arts, there is a substantial photographic section. Artists represented include British artist Damien Hirst, British sculptor Sarah Lucas, German photographer Wolfgang Tillmans, and American conceptual artist Jeff Koons, along with Danish artists Olafur Eliasson and Claus Carstensen.

Nautical Design

The irregular shapes of the spaces, the leaning walls, and the unpredictable layout all heighten the visitor's sense of curiosity and encourage exploration. At the heart of the design is a crescent-shaped gallery called the Art Axis. A sequence of further galleries, all lit by natural daylight, leads off this space. The Red Axis, a central passageway painted red, links all parts of the building and continues through the window of the Art Axis as a cement path, ending at a small jetty on the beach. The museum's restaurant hangs from the side of the building at level 2 like a lifeboat, its huge windows overlooking the sea.

SØREN ROBERT LUND

Born in 1962, Danish architect Søren Robert Lund was just 25, and still a student, when he won his first major commission to design the Arken Museum of Modern Art in the late 1980s. After leaving the Royal Academy of Fine Arts in Copenhagen, he established his practice in 1991. His buildings are strongly sculptural and expressive, and the landscape is always an integral part of the design process. His portfolio of projects includes private houses, corporate headquarters, schools, restaurants, a newspaper print works, and the Tivoli Hotel and Concert Hall in Denmark.

When offshore buoys transmit warning of a tidal surge, the gates take 1.5 hours to close.

The Delta Project's dams and sluices have shortened the coastline by almost 435 miles (700km).

In the past 50 years, sea levels around the Netherlands have risen by 9in (25cm).

Scheldt Tidal Barrier

With more than a quarter of the Netherlands lying below sea level, the Dutch have developed ingenious solutions to the danger posed by exceptional high tides. Nowhere is this more apparent than on the eastern outlet of the River Scheldt, where the world's largest tidal surge barrier protects Zeeland from the constant threat of flooding.

Completed in 1987 at a cost of US$3.1 billion, the barrier forms part of the Delta Project, a line of massive sea-defenses, which was constructed after disastrous floods in 1953 killed 1,835 people and destroyed more than 4,000 homes. As originally planned, the barrier was to be a solid dam that, by drastically restricting tidal flows, would have turned the Eastern Scheldt into a brackish lake and destroyed much of its marine environment. In 1973, however, a campaign forced the project to be suspended. The project was revived four years later in a new and greatly altered form. Instead of being permanently sealed, the redesigned barrier is now normally open to the tides, restricting their flow by only 25 percent. It is only when water levels reach 10ft (3m) above normal that steel gates, each 130ft (40m) wide, swing into position, holding back even the most extreme waters.

The eight-year construction program involved unprecedented work on a Herculean scale. The barrier's three sections contain 65 concrete piers, each almost 130ft (40m) high and weighing 20,160 tons (18,000 tonnes). These were cast on an artificial island that included, harbors and workyards, as well as a dry dock.

SHELLS BOATS

A fleet of custom-designed vessels was constructed for the project. The *Mytilos* (Mussel) held a floating gantry 165ft (50m) high from which giant needles 7ft (2m) in diameter and 60ft (18m) long were plunged into the sea-bed and vibrated at a frequency of 25–30 hertz. The *Cardium* (Cockle) carried a huge cylinder to hold the gravel-filled mats, which were 14in (36cm) thick. The most powerful ship was the *Ostria* (Oyster), a 285ft (87m) vessel with an 8,000 horsepower engine, designed to carry the 20,160-ton (18,000-tonne) piers. Other craft were also built for a variety of specialized tasks.

Pushing Skills to the Limits

Meanwhile, as the piers were being built, the sea-bed had to be prepared. First, huge vibrating needles were employed to consolidate the muddy sand by expelling excess moisture. Then, once the bottom had been leveled, gravel-packed mattresses were laid to form a solid base for the piers, which were carried from the dry dock on a massive floating hoist. Finally the 62 steel gates were suspended in position.

Built in turbulent waters up to 150ft (40m) deep, the barrier is increasingly being called upon to prove its worth as climate changes occur. In 1987 it was estimated that dangerously high tides would prompt the closure of its gates once every four or five years. In fact, in the first seven years of the barrier's operation it had to be closed on more than 20 occasions.

NOT THE ANSWER

Some conservationists continue to oppose the barrier's existence on the grounds of its environmental impact. Studies suggest that its effect on normal tidal flows has been greater than expected, causing silt accumulation that has harmed mussel beds and other marine habitats. It has also been suggested that such massive, interventionist schemes are not a sustainable response to global changes in sea levels, as sea levels are predicted to rise by up to 26in (67cm) by 2050.

Some Dutch architects have even put forward the possibility of floating cities as a long-term solution to the Netherlands' ever-worsening conflict with the sea, but the viability of this idea has not yet been proven.

41

750 passenger trains pass through the station every day at a rate of one every one-and-a-half minutes.

By 2010 an estimated 50 million passengers will be using the Lehrter Station every year.

Each individual module in the photovoltaic generator is a different shape.

Lehrter Station

The new Lehrter Station in Berlin, which was completed in 2006, occupies the site of an older one. The original Lehrter station was opened in 1871 in a bend of the River Spree, but was destroyed in World War II. The completion of the new Lehrter is another significant step toward the reassertion of Berlin as Germany's capital since the destruction of the Berlin Wall in 1989.

As well as helping to link together the disparate transport networks of the hitherto divided city, the station also links the country into the European high-speed rail network, connects local rail and underground services, and is a startling new state-of-the-art symbol for the city. Fully embracing modern environmentally-aware technology, the station's glass roof also acts as a carbon-free electricity generating station.

Solar Power

The station building, designed by the German firm Von Gerkan, Marg and Partner, stands at the intersection of the newly built north–south railway line, which runs in a tunnel 49ft (15m) below ground, passing below the River Spree and the Tiergarten, and the east–west ICE line, located 33ft (10m) above street level, enclosed in an elegant sleeve of barrel-vaulted glass. Two glass and steel towers rise up either side of the main concourse roof. Constructed with a grid of glass panels held in a lattice of cables and lightweight frames, this elliptical roof measures some 351 yards (321m) in length and spans six platforms.

Integrated into the south-facing side of the roof is a photovoltaic generator comprising 1,440 glass modules, which contain 133,200 high-efficiency mono-crystalline cells. The generator is linked to the national grid and provides electricity for the whole building.

The gross floor area of the entire project is 215sq yards (180,000sq m), including a platform area of 42,000sq yards (35,000sq m).

THE BERLIN WALL

During the Cold War in 1961 the Berlin Wall was built to provide a barrier between the east and west parts of the former German capital. In fact, the division of the metropolis happened much earlier, in 1945, when the Red Army captured Berlin at the end of World War II. When complete the wall was 96 miles (154km) long. It is estimated that there were about 5,000 successful escapees from the eastern side, but alongside these 192 were killed and 200 seriously injured in their attempts to flee. Eventually, as the Eastern Bloc crumbled, protesters breached the Berlin Wall on 9 November, 1989. Gleeful trophy hunters took chunks of the wall, and almost nothing of it is left today.

PHOTOVOLTAICS

Photovoltaics (PV), or solar cells, convert light directly into electricity. They are an inexpensive, pollution-free form of energy production. French physicist Edmund Bequerel first noticed the photoelectric effect in 1839, when he found that certain materials would produce small amounts of electric current when exposed to light. Photovoltaics are made with semiconductor materials, such as silicon, which are used in the microelectronics industry. A number of solar cells connected to each other and mounted in a support structure, or frame, is called a photovoltaic module. With the appropriate power conversion equipment, PV systems can produce alternating current (AC) compatible with any appliances, and be linked to the national grid.

Few buildings have to carry such a weight of symbolism as the historic seat of Germany's parliament in Berlin, which lay in ruins at the end of World War II. Now, with Germany reunified, the Reichstag has been reborn from the ashes of its past. Capped with an astounding dome of glass and steel, the old structure has been architecturally transformed into a shining beacon, sending out a symbolic message of democracy and peace.

Reichstag

n 1993 British architect Norman Foster won an international competition for the rebuilding of the Reichstag, although it was not until two years later that a final design was agreed upon. The challenge Foster faced was to accommodate three quite separate demands— the building had to function as an efficient working parliament that would meet the needs of members, the public, and the press. It also had to be environmentally efficient, consuming a minimum of energy despite its massive scale. Lastly, its architecture had to be expressive of modern Germany's ideals, looking optimistically toward the future without ignoring painful lessons from the past.

Completed in 1999, the building is a fascinating combination of the old and new. Intimidating, dark interiors have been swept away and replaced by transparent spaces filled with light and air. The glass-walled debating chamber allows the public unprecedented access to Bundestag debates, with seating for 400. The roof is also public space, with a terrace restaurant that is shared by both visitors and politicians.

Some fragments of the past have, however, been preserved. An aerial corridor of stainless steel and glass floats beneath an ornate vault that was carved in the age of the Kaisers. Doorways exhibit battered stone, and some walls are scrawled with graffiti in Cyrillic script, written by Red Army soldiers during the conquest of Berlin in 1945.

the facts

The total floor area is 73,200sq yards (61,166sq m).

More than 50,400 tons (45,000 tonnes) of demolition material was removed, amounting to a third of the building's fabric.

Hitler's architect, Albert Speer, planned to replace the Reichstag with an assembly housed in a 963ft (290m)-high dome.

In 1995 the American Bulgarian-born artist Christo wrapped the Reichstag in silver fabric to exorcise the spirits of its past.

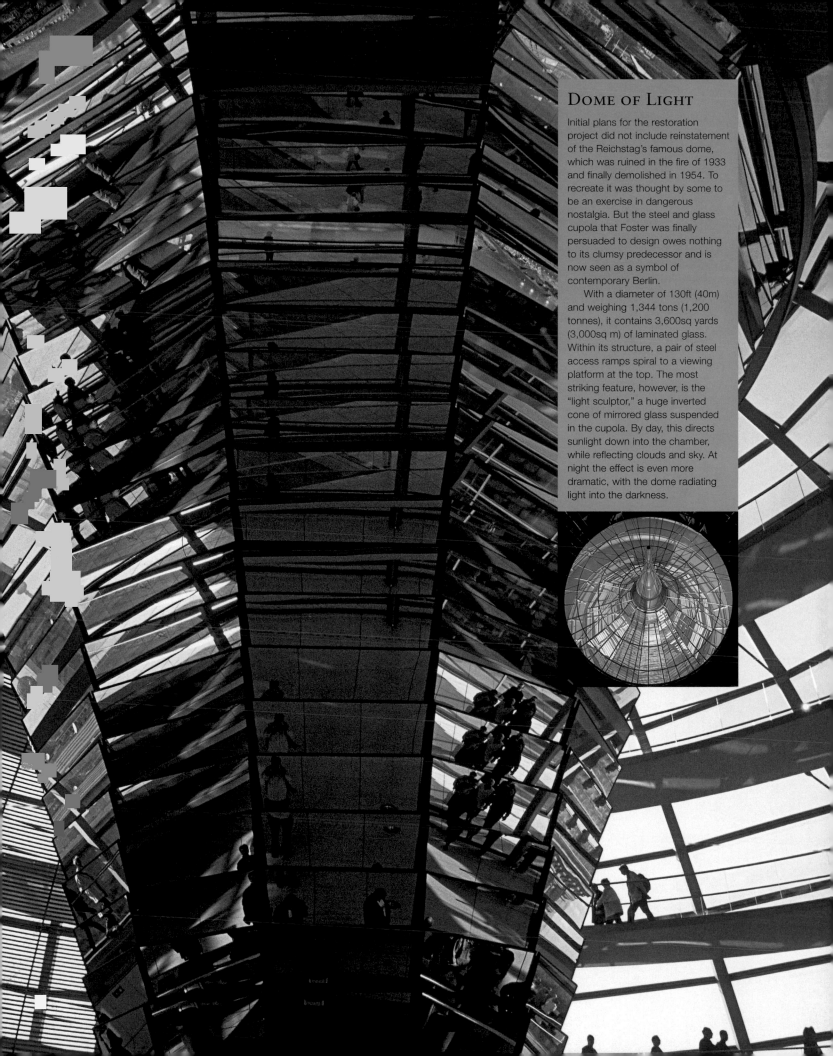

DOME OF LIGHT

Initial plans for the restoration project did not include reinstatement of the Reichstag's famous dome, which was ruined in the fire of 1933 and finally demolished in 1954. To recreate it was thought by some to be an exercise in dangerous nostalgia. But the steel and glass cupola that Foster was finally persuaded to design owes nothing to its clumsy predecessor and is now seen as a symbol of contemporary Berlin.

With a diameter of 130ft (40m) and weighing 1,344 tons (1,200 tonnes), it contains 3,600sq yards (3,000sq m) of laminated glass. Within its structure, a pair of steel access ramps spiral to a viewing platform at the top. The most striking feature, however, is the "light sculptor," a huge inverted cone of mirrored glass suspended in the cupola. By day, this directs sunlight down into the chamber, while reflecting clouds and sky. At night the effect is even more dramatic, with the dome radiating light into the darkness.

A History

The Reichstag, like the German nation, has endured a troubled history. Kaiser Wilhelm I laid the foundation stone in 1884, and it opened as a National Assembly in 1894. Following Germany's surrender in 1918, it became the debating chamber of the Weimar Republic, but its powers were curtailed when the Nazis rose to power. In February 1933 Dutch communist Marinus van der Lubbe set the building on fire, and the Red Army inflicted further massive damage in 1945. Crudely restored in the 1960s, when historic features were removed or concealed behind asbestos panels, it was occasionally used by government committees visiting Berlin. But it remained a focus for the dreams of those who hoped to see the city restored as the nation's capital.

These dreams were realized in 1989 when Soviet control over East Germany collapsed and the Berlin Wall was demolished. On 4 October, 1990, a day after Germany was reunified, the Bundestag held a symbolic meeting in the Reichstag, the first such assembly for 57 years.

Clean and Green

The Reichstag is one of the most environmentally efficient government buildings in the world. Aside from 359sq yards (300sq m) of solar panels on the roof, it has its own generating plant that produces "clean" electricity from rapeseed oil, a renewable resource. In summer, when demand is low, waste energy is stored by heating water that is discharged down into an acquifer reservoir 1,000ft (300m) below ground. At this depth the water suffers almost no heat loss, and can be pumped up in winter to underfloor pipes. A second aquifer, at 200ft (60m), stores cold water that can be used as a cooling system in hot weather. In the debating chamber, the "light sculptor" is also designed to perform a range of eco-friendly tasks. Not only does it light the chamber, using computerized, solar-powered sunshades to eliminate excessive glare, but it also releases stale air. Even the hot air generated in the chamber does not go to waste, as it is passed through heat exchangers before expulsion through the roof.

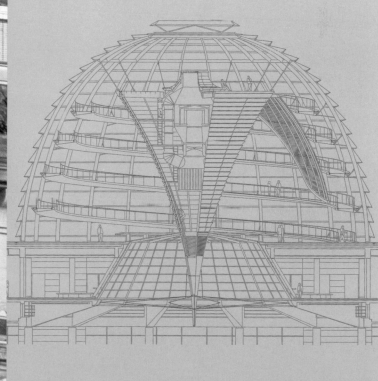

the **facts**

- The building is clad in untreated zinc that over time will oxidize to give a blue-grey sheen.

- The museum has a floor area of 18,000sq yards (15,000sq m), spread over four levels and a basement.

- In the Garden of Exile, 48 columns are filled with Berlin earth, while one contains soil from Jerusalem.

Jewish Museum

When a building receives international acclaim, it is often partly due to an elegant design that is pleasing to the eye. This is not the case with the Jewish Museum in Berlin. Designed to commemorate events too terrible to imagine, it is a building that defies the senses and the mind with strange proportions, uncomfortable surroundings, and a seemingly irrational plan.

Daniel Libeskind's design won a competition held in 1988 for a new annex to West Berlin's city museum that was to tell the history of Jewish communities in Germany from medieval origins to annihilation under Nazi rule. It is this last dreadful episode that informs the concept of a

JEWISH LIFE

The museum covers every aspect of Jewish life in Germany over the past 300 years, from religion and culture to politics. The collections include silverware and textiles, ceremonial and domestic objects, paintings and historic books. Along with a learning center that holds regular lectures and events, there is an important archive of community records, photographs, and private papers, including individual memoirs from the 1790s to the present day.

Above all, the museum is both a record and a celebration of the lives of the 200,000 Jews from Berlin who were killed or deported during Adolf Hitler's rule.

building that makes sense only as an abstract work of art. The project, which was fraught with disagreements over aims and problems with the budget, came close to cancellation following Berlin's reunification in 1990, but was finally completed by the end of 1998. During the 18 months before the collections were installed, 350,000 people toured the empty shell and spread the word about a building that is like no other.

A Dark Void

From the outside the building represents a brutally torn apart Star of David, a jagged zigzag reminiscent of a lightning bolt. Entry is through a tunnel from the old museum, a restored baroque palace dating from the 18th century. Inside bridges link galleries across an empty void and windows resemble vicious slashes cut into the walls. On emerging into the new museum you are confronted with three choices. Ahead, a long, straight staircase ascends to the main galleries, passing under concrete beams that seem to hold the walls apart. A second route leads to a dead end in the Holocaust Tower, a cold, dark cell with bare concrete walls that represents the fate of the millions sent to death camps. The third route leads symbolically to the outside world. Outside, the disturbingly unnatural Garden of Exile throws you off balance with its sloping ground, and tangled vegetation spilling down from columns instead of growing from the soil.

DANIEL LIBESKIND

When Daniel Libeskind won the museum competition, beating 165 rival entries, the 42-year-old architect had not designed a building that had actually been built. Libeskind was born in Poland in 1946 and studied music in Tel Aviv, and architecture in New York and England, before establishing a reputation for his writings and conceptual designs. The Jewish Museum catapulted him to international fame and he has since worked on major projects around the world. In 2003 he won the competition to design a replacement for the World Trade Center in New York.

The Marlene Dietrich Collection has 440 pairs of the star's shoes, 15,000 photos and 2,500 sound recordings.

The textile material of the Forum consists of self-cleaning Teflon fabric coated with glass fiber.

The fabric roof of the Forum is 112 yards (102m) wide at its widest point.

Sony Center

The Sony Center in Berlin is part of a vast new cultural and commercial quarter comprising seven buildings and covering some 119 acres (48ha) on Potsdamer Platz. A number of well-known multinational corporations including Sony and Daimler-Chrysler have been responsible for developing sections of the scheme. Along with its own offices, Sony has its own four-story flagship Style Store and Sony Professional Center on site.

Integrating Old and New

Sitting on a large, triangular wedge of land, the focus of the Sony Center, designed by Helmut Jahn of the Chicago practice Murphy/Jahn, is a vast elliptical, tented enclosure called the Forum. The high-tech canopy, constructed with tension wires and Teflon-coated fabric, looks like an open umbrella and encloses a great public plaza that serves as an arena for public events and performances. Sunlight filters through the canopy by day, and at night it is illuminated to create an impressive landmark. The lighting program begins with white light, but as dusk approaches the colors change to night-blue, cyan, and magenta. Framing the Forum are cafés, offices, and shops.

The complex is also a focus for the German film industry. It incorporates the Film Museum Berlin with its prized Marlene Dietrich Collection, numerous movie theaters including the state-of-the-art, eight-screen CineStar Original theater, and the CineStar IMAX movie theater with its 718-sq yard (600sq m) screen. The Potsdamer Platz complex is also the home of the German Film and Television Academy.

Another element of the complex is the Grand Hotel Esplanade, which first opened in 1908. Almost completely destroyed during World War II, it was declared a city landmark in 1989. In fact, one of its grand reception rooms, the Kaisersaal (a favorite of the last German Kaiser, Willhelm II), has been turned into a gourmet restaurant. Remarkably, to reintegrate the historic building into the new development, the room was jacked up and moved more than 80 yards (70m) to its current position.

HELMUT JAHN

Helmut Jahn was born in Nuremberg in 1940. He started his career in a Munich architecture practice, and at the age of 26 he moved to Chicago to join the prestigious offices of CF Murphy Associates, known for its work on a range of Chicago's most distinguished modernist corporate buildings.

Jahn is noted for combining technical and aesthetic expertise in handling steel and glass. His projects include the Frankfurt Messeturm (1991), which is a landmark in the German city and one of Europe's highest buildings. He has also designed the State of Illinois Center (1985), Chicago's United Airlines Terminal (1987), and the Shanghai New International Expo Center (2001).

POTSDAMER PLATZ

During the past century the story of Potsdamer Platz has been one of extremes. At the heart of Berlin, this central square was alive with activity in the early 20th century. It was a hub for the railroads, a busy traffic intersection for cars and trams (the first traffic lights outside of the United States were used here), it boasted some of the city's finest department stores, hotels, restaurants, and cafés, and was home to Berlin's famous nightlife.

During World War II Potsdamer Platz suffered devastating bomb damage, and then declined into wasteland when the Berlin Wall was built through its center in 1961. Today, with the fall of the Wall and reunification in 1989, Potsdamer Platz is once again at the center of the city, and its redevelopment made it the largest urban construction site in Europe at the turn of the millennium.

The building incorporates 4,200 windows with a total area of glass covering 36,000sq yards (30,000sq m).

The total amount of reinforced steel used in construction was 13,668 tons (12,400 tonnes).

There are 360 miles (580km) of electrical cabling in the building and 15 miles (24km) of sprinkler piping.

HISTORIC BANKING CENTER

Sometimes nicknamed Bankfurt, the city of Frankfurt is known internationally as an important global banking center. The banking tradition of Germany's fifth-largest city dates back to the Hanseatic League of the 14th century, and the money-lending empire of Jacob Fugger and family in the 15th and 16th centuries.

By the early 1800s Frankfurt was a thriving banking center under the Rothschild dynasty. Business grew quickly and by the late 19th century numerous new banks had opened. Today the city is home to 400 international and domestic banks, including the mighty German Bundesbank and the European Central Bank, which sets monetary policy for the Eurozone economy. The city's Stock Exchange is Germany's largest.

DZ Bank Tower

An unmistakable landmark on the Frankfurt skyline, the tallest element of the DZ Bank Headquarters building stands a full 52 stories. The tower, curved and clad in glass, is located in the southwest corner of the site to minimize the impact of its shadow on the residential Westend area, and to maximize its visual impact along the famous commercial corridor of Mainzer Landstrasse. At the top is an intriguing fan-shaped crown, which can be seen from streets around. It weighs 105 tons (95 tonnes) and is heated in the winter to prevent ice forming, which could dislodge and cause damage on the street below. The DZ Bank occupies the tower and has its own entrances.

There are two other elements to the building. The lower rectangular block, clad in stone, faces the residential neighborhood and encloses a public winter garden. The principal structural material is concrete, with finishes in glass, painted aluminum, painted steel, and honed granite. This, along with the third, lower still building, has a range of uses. In all, there are 15 apartments, office space, stores, and parking on three levels for more than 600 cars.

Keeping Cool

Along with high-level security features, including an area of secure parking with heightened surveillance and alarm systems specially for women, the buildings provide exceptionally comfortable workspace. There is natural daylight in all the offices (a requirement of German building regulations), impressive energy efficiency, and a raft of other ecological credentials, including triple glazing with low-reflective insulated glass on the exterior. Fabric roller-shades fitted between the layers of glass are unfurled automatically in summer to provide shade, and excess heat is drawn from between the layers of glass to maintain a comfortable temperature.

KOHN PEDERSEN FOX

Established in New York in 1976 by Eugene Kohn, William Pedersen, and Sheldon Fox, this practice, often known simply as KPF, has worked on a huge variety of buildings. Its impressive portfolio of completed projects ranges from Philadelphia International Airport and the World Bank Headquarters in Washington DC to the Shanghai World Financial Center in China. It has acquired almost cult status among fellow architects for its skyscraper schemes, confidently and expertly incorporating the latest design ideas, technological advances, and environmental considerations. The company now has offices in New York, London, and Shanghai.

the facts

The Commerzbank has a floor area of 144,000sq yards (120,000sq m) and houses 2,400 workstations.

It was Europe's tallest building until 2004, when it was overtaken by Moscow's 866ft (264m) Triumph Palace.

To reduce wastage of fresh water, restrooms use rainwater rather than the mains supply.

Commerzbank Tower

Frankfurt, along with London and Paris, is one of the few European cities to encourage high-rise building schemes. As Germany's main financial hub, the city cultivates a thrusting, modern image, with competing towers of steel and glass soaring high above the River Main. Germany is also a world leader in ecological reform, and businesses are expected to be environmentally responsible. As the world's first "ecological skyscraper," the Commerzbank Tower is a triumphant demonstration of how it is possible to reconcile such apparently conflicting requirements.

Located at the heart of the financial district, the building's 53 stories rise to 850ft (262m), with a mast on top taking its full height to 981ft (299m). Its distinctive triangular form, with facades of glass and aluminum, dominates the city skyline and is recognized worldwide. What is less apparent from outside is that this monument to company prestige also shows a sensitive response to ecological concerns. Norman Foster's revolutionary design places workers in close contact with a natural environment, while innovative measures have been introduced to reduce wasted energy and carbon-dioxide emissions. Since the completion of the tower in 1997, this "green" approach has influenced skyscraper design worldwide.

An Inside Out Tower

Conventionally, high-rise buildings are constructed around a core that provides structural support along with enough room for elevators and other services. However, at the heart of the Commerzbank there is nothing but empty space, a 49-story atrium that acts as the building's "stem," drawing in fresh air and light. The building's corner towers hold the floors, leaving the interior entirely free. Still more remarkably, a series of glass-enclosed gardens, each four

stories high, spirals up around the building's sides, occupying one third of its internal space. Every floor overlooks one of these nine gardens, and each garden is landscaped in its own individual style.

Another unusual feature is that windows can be opened manually. In the event of a sudden storm or high levels of air pollution, though, the building's own sophisticated system takes over. This system controls heating, ventilation, lighting, and even automated sunshades, constantly monitoring conditions for maximum energy efficiency. Occasionally the system can prove over-sensitive; if workers are inactive for too long, the lights around them automatically switch off. Overall, results are impressive, with energy costs reduced by two-thirds.

LORD FOSTER

In a career spanning more than 40 years, Norman Foster has become one of the world's most influential architects. Born in 1935, he had a working-class upbringing and paid his own way through his studies at Manchester University in England. Following a spell at Yale in the United States and a brief partnership with Richard Rogers, he founded his own London practice in 1967. Known for his imaginative use of steel, glass, and aluminum, along with an interest in environmental factors, he has worked with his team on projects in 48 countries, ranging from skyscrapers to bridges. In 1999 he was ennobled as Lord Foster of Thamesbank.

SKY GARDENS

The concept of "gardens in the sky" dates back to the days of Ancient Babylon, but it has acquired a new meaning in the age of skyscrapers. Malaysian architect Kenneth Yeang is a leading exponent of such schemes, with a proposed development of three "eco-towers" in London, while the Bionic Tower is a visionary project, complete with parks and gardens, by the Madrid-based practice Cervera & Pioz. As yet the Commerzbank is one of the few examples actually to be built. Gardens in high-rise buildings occupy valuable space and, unless protected behind glass, few plants thrive in these conditions.

The tunnel is lined with 1,232 superconductor magnets, which weigh 42 tons (35 tonnes) each.

Temperatures of −456°F (−271°C), colder than Outer Space, maintain the collider's superconductivity.

In 1996 scientists here successfully created atoms of antimatter, now routinely produced in experiments.

CERN HQ

One of the world's most astonishing laboratories lies hidden underground, burrowing beneath the frontier between Switzerland and France a few miles to the west of Geneva. CERN (the European Organization for Nuclear Research) was established in 1954 to provide Europe with a world-class center for advanced physics research. Now 6,500 scientists from 80 nations are involved in projects here that attempt to answer some of the last great mysteries of science through a new understanding of energy and matter. CERN's headquarters, based in the small Swiss town of Meyrin, provide some of

THE WORLD WIDE WEB

In 1989 Tim Berners-Lee, a British scientist at CERN, invented what was to become the World Wide Web. Combining the technologies of compatible computer systems, the Internet, and hypertext, the project was originally conceived as a global information system for scientists. The first browsers suitable for PC and Mackintosh were created at the University of Illinois in 1993, and by the end of 1994 the Web had 10,000 servers and 10 million users. CERN ceased to be involved in the Web's development in 1995, when the International Web Consortium was formed.

the most advanced facilities available for such research, and in 2007 one of the largest and most sophisticated machines that has ever been designed—the Large Hadron Collider (LHC)—will come into operation.

The LHC is a particle accelerator of unprecedented power, which fires beams of protons (electrically charged subatomic particles) in opposing directions around a 16-mile (27km) ring tunnel. When particles collide, traveling at close to the speed of light, matter splits from antimatter to produce a burst of energy.

Vast Caverns

The tunnel where experiments take place, and which is 2 miles (1km) longer than the Circle Line on London's Underground train network, extends beneath the foothills of the Jura Mountains. Constructed between 1983 and 1989, it originally housed a less sophisticated particle accelerator, which was decommissioned in 2000, but installation of the LHC has required additional engineering on a monumental scale. Experiments conducted with the new collider involve massive pieces of equipment such as the 13,780-ton (12,500-tonne) Compact Muon Solenoid (CMS) that generates a magnetic field 100,000 times stronger than that of earth, or the 82ft (25m)-diameter ATLAS detector that measures newly generated particles.

Each of these machines is housed in a cavern large enough to accommodate the dome of a cathedral, which was excavated from unstable rock that posed a constant challenge to the tunnelers. The ATLAS cavern has a concrete roof suspended from 38 steel cables anchored in galleries 82ft (25m) above. During construction of the CMS cavern, the engineers had to freeze the ground, using brine and liquid nitrogen, before they could dig access shafts down to the site.

FIELDS OF RESEARCH

While particle physics remains a largely abstract science concerned with the ultimate nature of the universe, there are many practical applications of CERN's research. Advanced forms of medical technology, such as the PET (Position Emission Tomography) scan, often used to detect cancer, have been developed from particle accelerators, as has mass spectrometry, used as a dating aid by archeologists.

Many new industrial processes and synthetic materials also derive from nuclear science. CERN has never been involved in weapons research, and although a gram of antimatter would release the same energy as a 40-kiloton nuclear bomb, the miniscule amounts produced at CERN are insufficient to pose any threat.

The bridge is 2,224ft (678m) long, with a main span of 570ft (174m). The tallest pier is 492ft (150m) high.

The central span is straight while the side spans are curved.

The piers flare outward above the road deck to give an added sense of substance and solidity.

Ganter Bridge

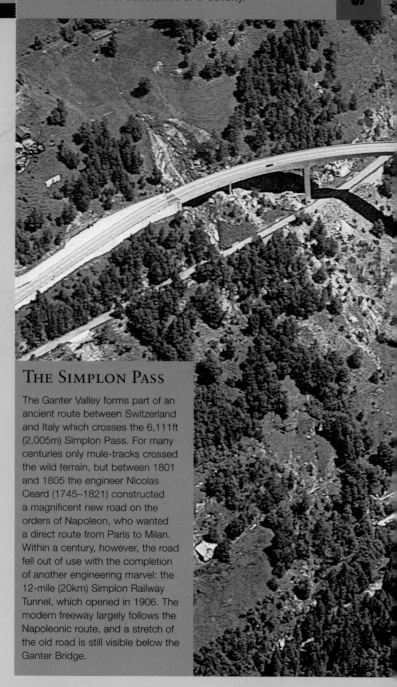

Many modern architects and engineers believe that "form should follow function," a saying that has sometimes justified structures which, although practical, show a disregard for popular ideas of beauty. The Swiss bridge-engineer Christian Menn saw the matter in a different light when he was asked to design a bridge across the Ganter Valley, in the Swiss Canton of Valais, in the 1970s. He agreed that the bridge should span the valley as a practical, cost-efficient structure. Given the extraordinary setting—a majestic Alpine landscape of forests, cliffs, and mountain peaks—Menn believed that the bridge also had to serve another function, of adding to, rather than detracting from, the beauty of the scene. The design that he produced lived up to this promise. Based on highly innovative engineering and construction methods, not only was the Ganter Bridge far cheaper than the tunnel that had originally been proposed, but it was also recognized as a masterpiece of architectural design. Menn had been a well-respected figure in his field for years, but the completion of this bridge in 1980 made him the world's most famous bridge designer.

THE SIMPLON PASS

The Ganter Valley forms part of an ancient route between Switzerland and Italy which crosses the 6,111ft (2,005m) Simplon Pass. For many centuries only mule-tracks crossed the wild terrain, but between 1801 and 1805 the engineer Nicolas Ceard (1745–1821) constructed a magnificent new road on the orders of Napoleon, who wanted a direct route from Paris to Milan. Within a century, however, the road fell out of use with the completion of another engineering marvel: the 12-mile (20km) Simplon Railway Tunnel, which opened in 1906. The modern freeway largely follows the Napoleonic route, and a stretch of the old road is still visible below the Ganter Bridge.

Style and Substance

In engineering terms the bridge is a brilliant hybrid based on two quite different methods of construction. The sturdy piers that soar up from the valley floor are not quite as solid as they look; they are effectively hollow boxes made from pre-stressed concrete. Comparatively light and yet with great lateral strength, they are designed to resist the stresses caused by sidewinds. The stylishly slender road deck is also a hollow-box construction, cantilevered out from the piers, which provide most of the support. The design so far is purely functional, but at this point true originality kicks in.

For a bridge this large to work as a cantilevered structure, it would either need to have a far bigger deck or the piers would have to be more closely spaced. Either choice would compromise the elegance of the design, so Menn explored a different option. Using the two central piers as supports, he added cable-stays for extra support. The steel cables are encased in concrete, which from a functional perspective gives them added strength. However, fossilizing the cables in cement also turned the bridge into a work of art.

MIRACLE MATERIAL

In 1849 a French gardener, Joseph Monier (1823–1906), discovered that by combining concrete with steel reinforcing bars he could produce giant tubs of unprecedented strength. Over subsequent decades engineers refined the principles of his discovery to perfect entirely new methods of construction that revolutionized both engineering and architectural design. By the start of the 20th century ETH Zurich, Switzerland's leading engineering college, was renowned for the brilliance of its teaching in this innovative field. Since then, many of the world's most visually astounding bridges have been designed by former students of the college, including Robert Maillart (1872–1940), Othmar Ammann (1879–1965), Heinz Isler (b. 1926), and Christian Menn (b. 1927).

Dr Michel Virlogeux, the designer of the bridge, later worked with Norman Foster on the Millau Viaduct.

The cable stays are composed of many steel strands, each strong enough to support 27 tons (24 tonnes).

The gap between the towers increases by half an inch between base and top due to the earth's curvature.

Normandy Bridge

When compared to other works of civil engineering, bridges have a glamorous appeal. They are, by their very nature, landmark structures that are used or seen by many people every day. Their existence can also be vital to the prosperous development of a region or a town. But bridges also have a mythical quality, seemingly defying the rules of nature as they span impossible divides. All these considerations played a part in the decision to construct a bridge of breathtaking scale across the River Seine between Le Havre and Honfleur, in northern France. Although traffic levels scarcely justified the cost of US$465 million, the bridge was seen as an investment in the future that would attract visitors to Normandy and be an example of masterful French engineering and style. With a clear gap of 2,808ft (856m), it had the longest cable-stayed bridge-span in the world at the time of its completion in 1995. Although it lost this title to Japan's Tatara Bridge in 1999, it is still the longest of its type in Europe.

Hidden Strength

Due to its scale and the difficulties posed by the site, the Normandy Bridge took almost seven years to build. The soft silt of the riverbanks could not provide the firm anchorage required for a conventional suspension bridge, so a cable-stayed structure was used. It is this system, with its elegantly angled network of supporting stays, which gives the huge bridge an appearance of delicate fragility that belies its actual size. The slender towers that support the road deck's cantilevered weight are 705ft (215m) tall, and each is made from 20,000 tons (18,000 tonnes) of pre-stressed concrete. The length of the bridge, including the two approach spans, is 7,025ft (2,141m) and its width of 75ft (23m) is sufficient for four lanes of traffic.

Strong and Light

Proven in tests to be able to support traffic weighing up to 16,000 tons (15,000 tonnes), and able to withstand winds of up to 185mph (300kph), the bridge is an extremely sturdy structure. It is testament to the skill of the designer, Michel Virlogeux, that the incredible strength of the bridge has been concealed within such a light and stylish form.

A Tale of Two Harbors

Separated by the broad estuary of the River Seine, Le Havre and Honfleur have developed in quite different ways. For centuries, both ports were equally important and of much the same size. Both suffered from blockades by the British navy during the Napoleonic wars.

But Le Havre soon recovered to become France's main Atlantic port, and despite being virtually destroyed during World War II, it is now a modern industrial city with a population of 300,000. Honfleur, by contrast, declined into obscurity as its harbor silted up, survived the war virtually unscathed, and is now a charmingly attractive seaside town with a population of only 8,000.

Cable-stayed Bridges

Cable-stayed bridges and suspension bridges are easily confused, but there are important differences in how they are designed. With a traditional suspension bridge such as San Francisco's Golden Gate, a pair of massive cables, spun on site from steel wires, is slung between the towers and secured to anchor points at each end. The deck is then suspended from smaller, vertical cables so that most of the weight is born by the anchors in the ground. With a cable-stayed bridge, the deck is suspended from precisely balanced fans of cables that are anchored only to the towers, which bear the structure's full weight.

National Library of France

The last of President Mitterrand's Grand Projets, the French National Library, in a prominent site on the Left Bank of the River Seine in Paris, is a striking building. Part of the reason for selecting this site between the bridges of Bercy and Tolbiac was to provide a focus for the regeneration of the area, and since the opening of the library, it has become a fixture on the tourist map. Inaugurated in 1995, it was designed by Dominique Perrault, who emphasized from the start that he was as interested in creating a place for people to use and enjoy as in designing a building.

Giant Books

The library comprises a quartet of L-shaped towers (resembling open books standing on end) placed at each corner of a central, low-level block surrounding a large open space. These four glass-sheathed, 260ft (79m)-high towers each accommodate seven floors of offices, all with movable wooden sun screens, and 11 storage levels protected by matching screens of insulating material. Initially the sheer walls of glass were criticized, as it was feared that the heat and light from the sun would damage the books.

The open esplanade, reached by a flight of wide steps from the Seine embankment, is a large public space decked with timber. A sunken garden covering 2 acres (1ha) sits in the middle of this, while the reading rooms occupy two levels around the garden, with workshops and stockrooms encircling them on the outer side. The belvedere in the northeast tower is open to the public. The book storage system comprises some 245 miles (395km) of shelving located partly within the central block, next to the reading rooms, and partly in the upper stories of the tower blocks. Service areas encircle the reading rooms at each level. You can reach the two library entrances by symmetrical, sloping walkways along the shorter sides of the garden on the east and west sides of the building. There are also reading rooms, an auditorium, a lecture hall, two exhibition areas, and a book store.

ROYAL COLLECTION

The founding of the National Library is attributed to Louis XI, who reigned from 1461 to 1483. His son, Charles VIII, followed by Louis XII, both added more early manuscripts along with printed books brought home from campaigns in Italy. An inventory of 1622 lists just 4,712 manuscripts and printed works. The collection expanded greatly under Louis XIV, and in 1720 the public was allowed access (once a week) for the first time. It has been estimated that more than 250,000 books were added during the revolutionary period. As the vast resource grew the library premises were expanded, first during the 19th century and again throughout the 20th century, until eventually the new library building was commissioned in 1989.

DOMINIQUE PERRAULT

Born in 1953, French architect Dominique Perrault trained in Paris at the École Nationale Supérieure des Beaux-Arts de Paris, where he graduated in 1978. Designing and building the French National Library was a huge challenge and responsibility for an architect who had not yet reached his 40s. Since the library, Perrault has gone on to work on a broad range of projects in Europe. These include the Olympic Velodrome and Swimming Pool in Berlin (1999), Innsbruck Town Hall in Austria (2002), and the European Court of Justice in Luxembourg.

- Changing art exhibitions are held at the top of the building in the rooftop galleries.
- The hole inside the arch is said to be large enough to accommodate the city's Notre-Dame Cathedral.
- In 1999 French climber Alain "Spiderman" Robert climbed the exterior wall of the arch without ropes.

Grande Arche

La Grande Arche is a huge modern "gateway," which stands in the modern business district of Paris. It is a modern interpretation of the Arc de Triomphe, the triumphal arch commissioned in 1806 by Napoleon to celebrate his military conquests, and completed in 1836. Measuring about 120 yards (110m) on each side, the Arche is a near-perfect cube. At the base is an intriguing tented reception structure known as the Cloud, and the glass-pod elevators. These elevators climb vertiginously to the underside of the arch top, whisking visitors up to the exhibition spaces and the rooftop viewing platform in just a few seconds.

Celebration of Victory

The Grande Arche de la Fraternité was completed in 1990 as one of the series of Grand Projets conceived by President François Mitterrand. It was designed to serve as a monument to mark the bicentenary of the storming of the Bastille and the 1789 French Revolution, and was inaugurated (when not quite complete) in July 1989 with grand military parades. The competition to design the project attracted more than 400 entries, and was won by the little-known Danish architect Johann-Otto von Spreckelsen on the grounds of his ideas of "purity and strength."

JOHANN-OTTO VON SPRECKELSEN

Before winning the 1982 competition to build the Grande Arche, architect Johann-Otto von Spreckelsen (1929–87) was virtually unknown outside his own country of Denmark. His buildings included a series of modernist churches such as Vangede (1974) and Stavnsholt Kirke (1981), both in the north of Copenhagen, and Stavnsholt, in Farum (1981). It is to the credit of the competition organizers that they had the confidence to commission this little-known designer. Von Spreckelsen died before witnessing the opening of his monumental project, which was carried through to completion by French architect Paul Andreu.

LA DÉFENSE

Following World War II, in the 1950s, the area around La Défense was marked out to become a new business district, and high-rise office buildings were constructed along a main avenue. In the 1980s the Grande Arche put the district on the tourist map.

The building stands in a grand plaza, surrounded by a series of intriguing modern structures and a vast open-air sculpture show, with colorful pieces by artists such as Spanish painter Joan Miró (1893–1983) and American sculptor Alexander Calder (1898–1976). Around 140,000 people work and 30,000 people live in La Défense, which is one of Europe's largest business centers.

Axo Historique

Spreckelsen's Grande Arche is constructed with a pre-stressed concrete frame and is clad in glass and 2,800 panels of white Carrara marble, imported from Italy, which cover 7 acres (3ha). Although the arch is a stunning landmark, it is in fact a giant office block, providing around 108,000sq yards (90,000sq m) of office space over 35 floors for a workforce of more than 2,000 employees.

The arch stands in the chic, skyscraper-crammed business district of La Défense, and marks the western end of the vast *Axe historique* (historical axis) slicing through Paris—a line of monuments, buildings, and avenues running westward from the center of the city. The axis began with the Champs-Elysées in the 17th century and then expanded to include the Tuilleries Gardens and the Arc de Triomphe.

Louvre Pyramid

When plans to build a crystalline glass pyramid at the entrance to the Louvre Museum in Paris were first unveiled in the early 1980s, they were greeted with outrage. Many believed it was a travesty to build such an uncompromisingly modern structure in front of the grand and historic Louvre building. The massive museum edifice, built on the site of an ancient fortress, was begun in 1546 during the reign of François I, and stood as the royal palace of French kings until the late 17th century.

In 1793 it opened as a public museum, and by the 1980s the building had become congested, with one wing used as government offices and the front used as a car park. As part of his Grand Projets, President Mitterrand requested that the monument be expanded.

GRAND PROJETS

The Louvre Pyramid stands as the best known of the Paris Grand Projets. These were carried out during the 1980s at the instigation of President François Mitterrand (1919–96) to provide a series of modern monuments to symbolize France's central role in art, politics, and world economy at the end of the 20th century. Other landmark buildings include the marble-clad Grande Arche de la Défense, the glassy towers of the National Library by Dominique Perrault, and Carlos Ott's Opéra Bastille, a stepped glass building opened in 1989 in recognition of the bicentenary of the 1789 Revolution.

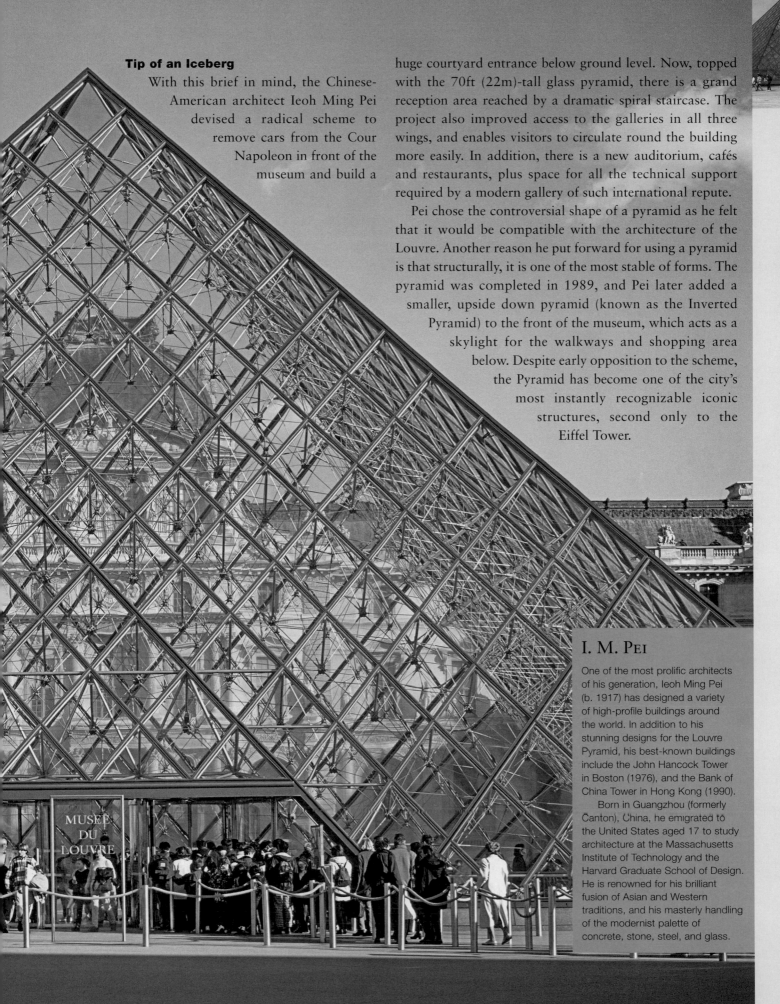

Tip of an Iceberg

With this brief in mind, the Chinese-American architect Ieoh Ming Pei devised a radical scheme to remove cars from the Cour Napoleon in front of the museum and build a huge courtyard entrance below ground level. Now, topped with the 70ft (22m)-tall glass pyramid, there is a grand reception area reached by a dramatic spiral staircase. The project also improved access to the galleries in all three wings, and enables visitors to circulate round the building more easily. In addition, there is a new auditorium, cafés and restaurants, plus space for all the technical support required by a modern gallery of such international repute.

Pei chose the controversial shape of a pyramid as he felt that it would be compatible with the architecture of the Louvre. Another reason he put forward for using a pyramid is that structurally, it is one of the most stable of forms. The pyramid was completed in 1989, and Pei later added a smaller, upside down pyramid (known as the Inverted Pyramid) to the front of the museum, which acts as a skylight for the walkways and shopping area below. Despite early opposition to the scheme, the Pyramid has become one of the city's most instantly recognizable iconic structures, second only to the Eiffel Tower.

I. M. PEI

One of the most prolific architects of his generation, Ieoh Ming Pei (b. 1917) has designed a variety of high-profile buildings around the world. In addition to his stunning designs for the Louvre Pyramid, his best-known buildings include the John Hancock Tower in Boston (1976), and the Bank of China Tower in Hong Kong (1990).

Born in Guangzhou (formerly Canton), China, he emigrated to the United States aged 17 to study architecture at the Massachusetts Institute of Technology and the Harvard Graduate School of Design. He is renowned for his brilliant fusion of Asian and Western traditions, and his masterly handling of the modernist palette of concrete, stone, steel, and glass.

the facts

- The tunnel was designed to carry 450,000 vehicles a year, but traffic has increased by 350 percent.

- The tunnel slopes gently from the Italian side at 4,487ft (1,368m) to the French side at 4,179ft (1,274m).

- A workforce of 235, including 64 firefighters and a 40-strong police force, is employed by the tunnel.

Mont Blanc Tunnel

The French town of Chamonix and the Italian town of Courmayeur are less than 8 miles (13km) apart on any map, but until recent times there was no easy route between the two. Roads took long detours through Alpine passes to the east or west, while walkers faced a three-day hike across the central massif of Mont Blanc, Western Europe's highest mountain. Although the first proposals for a tunnel were put forward in the 19th century, it was not until after World War II that the French and Italian governments recognized the feasibility of such a project. Following a promising test-bore on the Italian side in 1949, an agreement to proceed was signed later that year—although it was another 10 years before construction finally began.

The entire Mont Blanc Tunnel project challenged engineers to the limits of their skills. The 7-mile (12km) road tunnel was the longest that had ever been attempted, and in the early 21st century still remains by far the deepest tunnel in the world, running 8,134ft (2,480m) below the ground as it pierces the core of the mountain.

Difficult Conditions

Using engineering methods that had never previously been tried on such a scale, the tunnel was excavated by a giant tunnelling machine that advanced at the rate of up to 26ft (8m) a day. Work was often interrupted by disasters, including floods and roof collapses, while conditions on the surface were sometimes scarcely better in the harsh, high-altitude environment. On one occasion the Italian work-camp was virtually obliterated by an avalanche. Despite these set-backs, the two teams met in the center of the mountain on 14 August, 1962, with an off-set of less than 5in (13cm) between the alignments of their tunnels.

LESSONS LEARNT

On 24 March, 1999, a Belgian lorry caught fire 3 miles (5km) inside the tunnel, starting an inferno that raged uncontrolled for 53 hours and claimed 39 lives. Over the following three years the tunnel remained closed while it was given a major refurbishment. Safety features now include a heat-detector cable, new smoke-extractor flues, video surveillance, warning lights, and four water reservoirs. There are also 37 pressurized shelters linked to an escape route that runs beneath the carriageway. In addition to the main control room outside the French portal, there is a duplicate back-up on the Italian side in case of communications being broken.

THE MOUNTAIN

The height of Mont Blanc is generally given as 15,771ft (4,807m), although the top 75ft (23m) of this is solid ice that can vary in thickness from year to year. Michel Paccard and Jacques Balmat made the first ascent in 1786, an achievement that was then considered so remarkable that they were not universally believed. Now more than 2,000 people climb the mountain every year, while tens of thousands more visit its most famous glacier, the Mer de Glace. The easy access provided by the tunnel has undoubtedly put pressure on Mont Blanc's environment, although its wild slopes remain untamed.

The Final Phase

Once the excavation was finished and millions of tons of rock removed, the carriageway was then built, and a complex infrastructure that included ventilation ducts and powerful fans to expel fumes, was installed. This took three years, and the tunnel was finally opened on 16 July, 1965.

The bridge took a workforce of more than 400 just over three years to build.

The piers are made of 206,000 tons (188,000 tonnes) of concrete and 19,000 tons (17,000 tonnes) of steel.

The split piers are designed to flex as the road deck expands in the sun.

Millau Viaduct

The opening of the Millau Viaduct in December 2004 was greeted with a fanfare of international acclaim. Praise was lavished on the bridge both as a technical achievement and as a masterpiece of architectural design. But it was the sheer mind-boggling size of the structure that astonished most observers.

Carrying the A75 freeway 2 miles (3km) across the deep valley of the River Tarn, in central France, the viaduct was the world's highest elevated road when it opened. At its deepest point, the valley floor lies 886ft (270m) below the four-lane freeway, giving drivers an experience similar to flying a plane. When viewed from ground level, the scale of the bridge is equally phenomenal. The seven concrete piers that march across the valley are the size of skyscrapers—the tallest being 804ft (245m) high—while the steel masts supporting the suspension cables are each a further 295ft (90m) tall. From pier base to masthead, the bridge is taller than the Eiffel Tower and a close rival to the Empire State Building. But while such statistics are impressive, they give no indication of the viaduct's extraordinarily graceful form. Despite its monumental size, it seems to sail across the valley with the lightness of a racing yacht.

British architect Norman Foster and French bridge engineer Michel Virlogeux were equally involved in the design of the viaduct. In terms of engineering, it consists of eight independently supported cable-stayed spans, two of 222 yards (204m) and six of 370 yards (342m), yet the viaduct's beauty and strength are what make it so special.

SLIDING DECK

Given the enormous heights involved, the steel-box sections of the road deck could not be lifted into place with cranes. Instead, they were assembled on abutments and pushed into position from the sides. Seven temporary towers were built, halving the length of unsupported spans. Guided by GPS systems, hydraulic rams fastened to the pierheads then slowly edged the deck along its course. With the deck advancing in a series of four-minute, 2ft (0.6m) thrusts, it took three days to bridge each gap. Once this operation was completed, the masts and stays were moved into position from the road.

OVER OR UNDER

Fourteen years of planning and discussions preceded the construction of the bridge, which is part of a new freeway between Paris and Barcelona. The route across the valley—one of four originally considered—was chosen largely on the basis of environmental impact in relation to the town of Millau and adjacent villages. There was then a choice between the viaduct and a lower-level crossing that would have required the construction of a tunnel. Once this decision had been made, four separate teams of architects and engineers were invited to submit designs. The winner was announced in 1996, and construction work began in October 2001.

The Height of Beauty

What gives the viaduct its sublime elegance is the way in which this basic structure has been developed and refined in response to both structural requirements and aesthetic ideas. The colossal piers taper and divide in two below the deck, lightening their form. The masts are also split, with legs delicately balanced on the arms of piers, while the cable-stays fan down like sails fastened to the road-deck's central reservation. For the benefit of drivers, the road slopes at 3 percent from south to north and follows a slight curve, allowing the whole structure to be seen from any point along its length. Millau was once notorious as a bottleneck on the old road to the south, but now the bridge can be crossed in just 90 seconds and it is an experience that few travelers forget.

the **facts**

The museum's construction costs of $100 million were met by the Basque government.

30,200sq yards (25,220sq m) of titanium and 45,000 cubic yards (34,400cu m) of limestone were used.

As a precaution against the river flooding, the museum is anchored to bedrock 46ft (14m) below ground.

Guggenheim Museum

The creation of an architectural masterpiece can sometimes transform the reputation of an entire city in the eyes of the world. This was certainly so with Bilbao, an industrial port in northern Spain that was in a sad state of decline before the opening of Frank Gehry's Guggenheim Museum in 1997.

Within a year of its completion, this mind-boggling building of titanium-clad forms became the most talked-about example of modern architecture in the world and attracted two million visitors to Bilbao. Some detect within its wildly abstract shapes the image of a ship, while others see a scaly fish—a form much loved by Gehry. But whatever it may represent, it is a structure of astonishing originality that breaks all accepted rules of both traditional and modern architectural design. It even seemingly defies the limitations of the site on which it stands, pushing out over the waters of the River Nervion, and sliding an immense Long Gallery beneath a busy road-bridge to connect with a tower, apparently splitting open from within, that rises on the far side.

Method in the Madness

At first glance, the building looks like a mad collage of impossible geometry, with curving walls, asymmetric roofs, and cantilevered overhangs. The inside of the museum is in many ways as spectacular as its exterior. Beyond the entrance hall catwalks cross a soaring atrium, with glass elevator shafts rising to 165ft (50m). Rooms with bowing, tilting walls throw out of alignment all perceived conceptions of the world. Like the great cathedrals of the Gothic age, the museum is designed to inspire and captivate the senses. But practical internal spaces lie behind the seeming chaos of this outward form. Vertical, rectangular walls of polished stone front 10 galleries of

BUILT TO LAST

The museum's curvaceous forms are constructed from a complex frame of steel tubes bolted to an underlying skeleton of conventional rectilinear sections. The use of universal joints allowed minute adjustment of this frame in all directions prior to cladding with a 0.079in (2mm) layer of galvanized steel. The outer skin consists of titanium plates just 0.015in (0.38mm) thick. Titanium, a metallic element mined only in Russia and Australia, was chosen in preference to stainless steel because of its capacity to take on the color of ambient light. Despite being thin enough to ripple in the wind, the panels have a life expectancy of at least 100 years.

conventional design in which the permanent collection is housed. Nine further galleries, behind the curvaceous titanium facades, offer more intriguing spaces for visiting exhibitions that explore the wilder shores of modern art. Then there is the Long Gallery, which, at 425ft (130m) long by 97ft (30m) wide, is designed for artworks so colossal that they could not be exhibited elsewhere. In all, there are 13,000sq yards (11,000sq m) of exhibition space, amounting to almost half of the building's total floor area.

THE GUGGENHEIMS

Solomon R. Guggenheim (1861–1949) was a wealthy American industrialist who married Irene Rothschild in 1895. Up to the age of 68 his interest in art was limited to Old Master paintings, but in 1929 the German artist Hilla Rebay introduced him and his wife to modernism. With Rebay's help, the couple subsequently accumulated one of the world's finest collections of avant-garde art, which since 1959 has been based in New York's Guggenheim Museum, designed by Frank Lloyd Wright. In recent years the Guggenheim Foundation has expanded far beyond Manhattan, with outposts in Las Vegas, Venice, Berlin and, most notably, Bilbao.

The beaches surrounding the hotel are manmade, and sand is imported regularly to keep them looking good.

The Fish ("Peix"), 115ft (35m) tall and 177ft (54m) long, is visible from anywhere on the beach.

The hotel's grand ballroom can accommodate up to 1,000 guests.

Hotel Arts

Most people associate Barcelona with the amazing organic buildings of maverick architect Antoni Gaudí (1852–1926), the busy shopping and promenading street of las Ramblas, or the pulsating nightlife. Barcelona as a seaside town is altogether a more unusual image, despite the fact that it has a Mediterranean shore.

However, when the Olympic Games came to the city in 1992 a huge area of waterside industrial wasteland became the focus of regeneration and redevelopment. Olympic village accommodation and a new quarter with hotels, offices, homes, shopping, restaurants, a marina, and a manmade beach were created. This bold development succeeded in bringing the city to meet the sea, and a good transport system links the two; alternatively, it is little more than a 15-minute walk to las Ramblas.

1992 OLYMPICS

From the start of the opening celebrations, it was clear that the 1992 Barcelona Olympics would be an exciting and vibrant spectacle. Along with a roll call of more than 9,300 athletes from 169 nations, Barcelona enlisted the help of a cast of internationally renowned architects to create a series of amazing venues. Among the legacy of new buildings are the Sant Jordi Sports Hall, designed by Japanese architect Arata Isozaki, the 656ft (200m)-high communications tower by Britain's Foster and Partners, and a sculptural, sky-piercing Olympics monument by Spaniard Santiago Calatrava.

BIG FISH

The public space at the base of the hotel tower is dominated by Frank Gehry's huge fish sculpture. Originally he installed a trellis to act as a shade over the public courtyard. The trellis took the form of a wave and then became the abstract flying fish that can be seen today. The sculpture is made from the same exoskeleton as the hotel, but is clad with a different material to separate it from the rest of the building. Gehry's idea was that the fish would have a significant presence but that the tower would always remain dominant.

A Striking Landmark

Taking pride of place on Barcelona's new-look waterfront is the Hotel Arts, a stunning 44-story tower rising 502ft (153m) into the Spanish sky—a landmark for miles around in this largely low-rise city. The building shares the harborfront skyline with one of the city's other rare skyscrapers—the Mapfre Tower, which is used for office accommodation. When they were completed these were the tallest towers in Spain.

Designed by international practice Skidmore Owings & Merrill, the Hotel Arts cuts a graceful figure with its distinctive "exoskeleton." Much of the structure is worn on the outside of the building creating an intriguing steel-grid facade. Not surprisingly, the breathtaking views are a major selling point of this 600-room hotel. But there's plenty more on offer too, including luxuriously appointed contemporary-style interiors with sculptures, paintings, and an array of other artworks on display in the public areas and in guest rooms. There is also an impressive spa on the 43rd floor. Along with its striking architecture, the building is distinguished by the massive, gleaming, fish-shaped sculpture at its base, designed by the American architect Frank Gehry.

Expo '98 had the theme of The Ocean and attracted more than 11 million visitors during its 132-day run.

Over 140 countries and international organizations participated in Expo '98.

The Vasco da Gama Bridge can withstand winds of 155mph (350kph).

VASCO DA GAMA

In 1498 the Portuguese explorer Vasco da Gama made the first sea passage from Portugal to India, and 500 years later the cable-stayed Vasco da Gama Bridge opened in Lisbon as part of the Park of Nations project. At the time it was Europe's longest bridge being 10 miles (16km) from shore to shore. The bridge was designed to withstand an earthquake four times greater than the one that devastated Lisbon in 1775.

Oriente Station

With its graceful sweeping arches and crowning glass canopy, the Oriente Station is a striking addition to Lisbon's cityscape. Built for the Expo 1998 World Fair, it was designed by the Spanish architect and engineer Santiago Calatrava. Taking his inspiration from nature, Calatrava's highly expressive work is characterized by the use of plant and animal forms—his TGV station at Lyon in France, for example, has been compared to a vast prehistoric bird about to take flight. His bridges often have the poise and balance of great skeletal structures, and in every corner of this station in Lisbon you can glimpse shapes reminiscent of trees, joints, and bones.

A Network Hub

Expo '98 took place on what was previously industrial wasteland about 3 miles (5km) to the northeast of the city center, and its lasting legacy has been the continued regeneration of the area. This urban project, called the Park

SANTIAGO CALATRAVA

Architect, engineer, and artist Santiago Calatrava was born in 1951 near Valencia, in Spain. His background in architecture and engineering has fused with a passionate interest in nature to produce intriguing and elegant organic structures. Calatrava first came to fame with his beautiful bridges—the designs now number more than 50—which can be found in cities from Bilbao to Berlin, and from London to Toronto. Other eyecatching schemes include the TGV train station at Lyon in France (1994), Bilbao's Sondica Airport in Spain (2000), the Opera House on the Spanish island of Tenerife (2002), and the City of Arts and Sciences—an urban recreation center in his home town of Valencia in Spain.

of Nations, occupies a 340-acre (138ha) site with 3 miles (5km) of land fronting the River Tagus. The Oriente Station was vital in connecting the new neighborhood to the city center. The purpose of the station was to create a major transportation interchange where passengers could move from long- and medium-haul regional and international trains to metropolitan services and underground lines, buses, and taxis. An airport link and check-in were also incorporated.

When the station was built it had to accommodate the existing railway lines that crossed the district 30ft (9m) above street level, and Calatrava placed the four train platforms on a structure comprised of five parallel rows of twinned arches. The platforms shelter under the delicate tracery of a glass and steel roof, raised on pairs of steel pillars shaped like huge palm trees.

The platforms are reached by ramps or cylindrical glass elevators from the multilevel concourse below, where there are connecting routes to the underground railway and other forms of transport, ticket counters, and a shopping mall. Here you can see Santiago Calatrava's trademark use of poured concrete to create huge spans and arches not unlike the skeleton of a dinosaur. The bus terminal is immediately to the west of the station, and a complex of commercial buildings is arranged around a plaza to the east.

William Herschel Telescope

Europe's most extensive collection of large telescopes stands on the crest of a spectacular volcanic ridge 7,850ft (2,393m) above sea level on the island of La Palma. The island, which is one of the Canary Islands, lies a few hundred miles off the northwest coast of Africa. The high altitude and skies free from light pollution provide some of the best conditions for astronomical observation in the world. The conditions are so good that in 1979 Britain's Royal Greenwich Observatory settled on this site for its main observatory in the northern hemisphere.

In 1985 work began on the construction of the William Herschel Telescope, an immensely powerful instrument with a 13.8ft (4.2m)-diameter primary mirror that remains one of the largest examples of its type in the world. The project took more than two years to complete, and involved nerve-wracking challenges as huge parts of the delicate assembly were slowly maneuvered around steep hairpin bends. The distinctive onion dome that houses the great telescope is a highly sophisticated structure. Weighing 350 tons (320 tonnes), it rotates smoothly around a circular track above the observatory floor.

CLOSE ENCOUNTER

In August 2004 astronomers used advanced optics systems incorporated in the Herschel telescope to obtain remarkable pictures of an asteroid that passed within 500,000 miles (750,000km) of earth. Such relatively close encounters occur only once or twice a century, and while the risk of collision is low, it has occurred in the distant past with catastrophic consequences. In this first-ever observation of a Near Earth Asteroid, it was established that 2002NY40 was an elongated object 1,300ft (400m) long, tumbling through space at 40,000mph (65,000kph). However, approaching no closer than twice the distance from the Moon, it was judged to pose no threat.

Absolute Precision

The telescope's most vital element is its primary mirror that reflects and focuses the light from distant stars, often tracking targets over many hours to produce a clear, bright image. The parabolic disc is cast in Cervit, a glass-ceramic material that doesn't expand in changing temperatures. Since even the tiniest distortion of the mirror's surface would affect the sharpness of the image it reflects, extraordinary measures have been taken to ensure maximum precision. As the mirror tilts to follow targets up from the horizon, 60 pneumatic cushions, each with its own pressure sensor, constantly expand or contract to maintain all stresses at an even level. Even human activity within the dome is strictly limited in order to minimize air disturbance and vibrations. The control room, along with the computer systems, dark rooms, offices, and workshops, are all housed in a three-story annex.

MAGNIFICENT THREE

The three telescopes established on La Palma by the Royal Greenwich Observatory of England, are now known as the Isaac Newton Group. They form part of the Roque de los Muchachos Observatory, the most extensive astronomical facility in the northern hemisphere, with 12 telescopes of various types stretching along the ridge over nearly a square mile (2sq km). The Observatory is run under Spanish jurisdiction, although astronomers representing several different countries work here. A new super-powerful telescope, the Gran Telescopio Canarias, was completed in 2006.

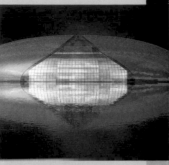

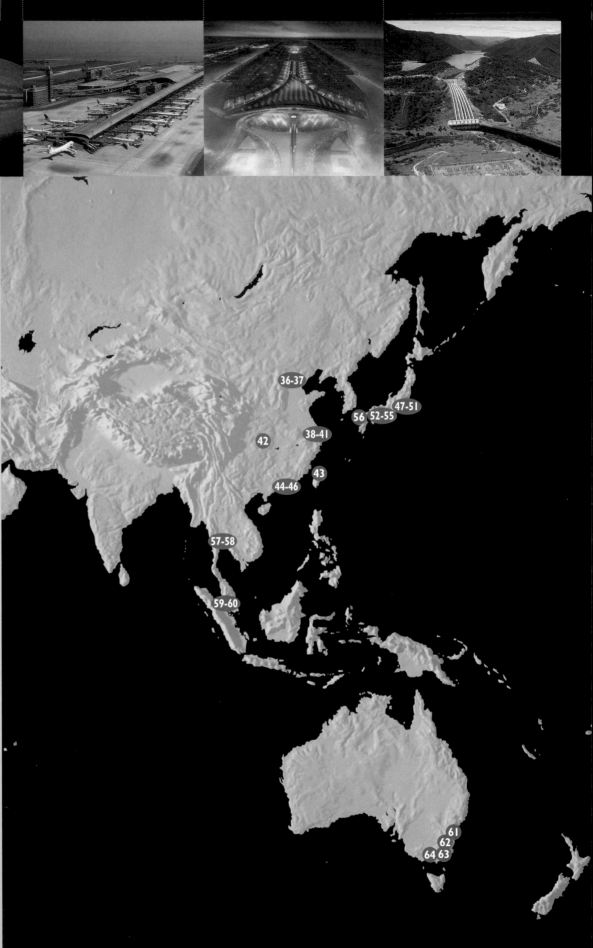

Asia and Australia

Always at the forefront of technology and innovation, Asia is still a frontrunner when it comes to dazzling, futuristic skyscrapers, high-speed trains, and mega building projects. As countries in Asia develop and their populations increase, they need increasingly sophisticated infrastructures, which has led to the construction of many new road bridges, and port and airport terminals, which will be able to cope with the increase in the volume of traffic. However, rather than simply creating functional, but ultimately unattractive constructions, many countries are employing architects known for their cutting-edge designs to create functional buildings where form is as much a part of the design as function. An example of this is Yokohama Port Terminal in Japan, where beautiful Brazilian hardwood has been used to create appealing, calming, waterfront walkways, which don't detract from the terminal's ability to cope with the passengers and cargo of four huge cruise liners at a time.

Asia is also a continent which likes to think big. Witness Malaysia's spectacular (and vertigo-inducing) Petronas Towers, whose twin towers dominate the skyline of Kuala Lumpur. The towers are so tall that they held the title of world's tallest buildings from 1996 until 2004, when they were surpassed by another Asian mega structure, the Taipei 101 building in Taiwan. However, the most awe-inspiring construction project of all (and one which has faced harsh criticism from environmentalists), is China's Three Gorges Dam. The dam is probably the largest construction project that has ever been undertaken, anywhere in the world. It involves a budget of $25 billion, a workforce of nearly 30,000, and the resettlement of more than a million people.

the facts

- Terminal 3 will be the most technologically advanced airport terminal in the world.
- To meet the 2007 deadline, construction teams will work non-stop around the clock until completion.
- Daylight will enter through the roof, providing natural light throughout the building.

Beijing Airport Terminal 3

Since its entry into the World Trade Organization in 2001, China's economy is now the fastest-growing in the world. This, and the influx of visitors to the 2008 Olympics, means the volume of passengers passing through Beijing's Capital Airport will have increased from 27 million in 2005 to around 60 million by 2015. To accommodate this massive expansion, work on the new Capital Airport Terminal 3 was started in March 2004, with a completion date set for 31 December, 2007. Foster and Partners won the contract to build the new terminal in November 2003.

The building of the terminal has two objectives. To more than double passenger capacity efficiently and safely, and to establish the airport as a symbol both of New China and of traditional Chinese national culture. The new twin passenger terminal buildings will have a roof area of more than 80 acres (32ha) and measure 880 yards (800m) across at the widest point. This means it will have a larger surface area than all five of London Heathrow's terminals put

FIGHTING THE SANDS

Located to the northeast of the capital, Beijing Airport is menaced by the increased intensity of dust storms blowing in from the Gobi Desert and Mongolian steppe. While drought and climate change are partly to blame, the root cause is overgrazing. Throughout China, more than 38,000sq miles (100,000sq km) of land has been turned into desert over the past half century as vegetation has been stripped away. By 1980, desertification northwest of Beijing exceeded 4 percent a year. To resolve this there has been massive reforestation, and the planting of an anti-desert "green belt" north of the capital, which will act as a barrier against the Gobi sands.

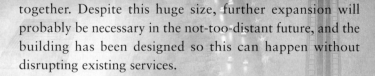

CELEBRITY ARCHITECTS

Beijing's construction boom aimed at preparing the city for the 2008 Olympic Games has provided exciting opportunities for celebrity architects, both from China and abroad. At a time when Western nations have been cutting back on construction expenses, China is spending freely on prestige buildings. Swiss architects Jacques Herzog and Pierre de Meuron have been awarded the 2008 National Stadium contract, and Paul Andreu designed the Beijing National Grand Theater.

together. Despite this huge size, further expansion will probably be necessary in the not-too-distant future, and the building has been designed so this can happen without disrupting existing services.

A Symbol of National Identity

An overriding priority is to make the airport immediately identifiable with China. At night, the glow from the airport's orange, yellow, and red lights will be visible through the massive roof, outlining the terminal's distinctive silhouette and suggesting the colors of the Chinese national flag. The roof has an uneven texture that has been likened to the scales on a dragon's back, and the architects consulted traditional Chinese geomancers on the *feng shui* of the area, designing the arrivals area in particular to be both calming and inviting to passengers.

The terminal has also been designed to be one of the world's most sustainable airports. It incorporates southeast-orientated skylights to maximize warmth from the early-morning sun, as well as integrated environmental control systems to minimize the consumption of energy and carbon emissions.

52 elevators and 36 escalators serve the interior of the theater.

A 196ft (60m)-long transparent underwater passage links the main entrance with the reception lobby.

In the event of an emergency, the audience can escape within six minutes.

the facts

National Grand Theater

Beijing's National Grand Theater is a high-profile example of the stylish, international architecture that the Chinese authorities are adopting in and around their 800-year-old capital, Beijing, in time for the 2008 Olympic Games. The creation of French architect Paul Andreu, the design of the National Grand Theater is deliberately designed to be iconic, symbolizing the country's increasingly influential and affluent status in the world. Like the Louvre Pyramid in Paris, or the Sydney Opera House, it is hoped that the building will change perceptions of traditional Beijing, both in the eyes of its inhabitants and of the outside world. Budgeted at a cost of US$335 million, the theater is a vast, domed, 21-story structure which can seat several thousand

people. The "bubble," or "egg," as it has become known, is made of 1,200 pieces of glass and 20,000 titanium plates, and appears to float on an artificial lake. Its rounded surface covers 215,300sq yards (180,000sq m).

Inside are three huge performing halls: a 2,400-seat opera stage in the center; a 2,000-seat concert hall to the east; and a 1,500-seat theater to the west. There will also be exhibition halls and an audio and video store. An underground car park will have space for more than 1,000 cars and 1,400 bicycles, and there will be direct access to the subway to ease congestion after performances have ended.

Controversial Location

Impressive as the theater is, many local people feel it would have been better placed in a newer part of the city, well away from the highly respected Forbidden City. It stands opposite the Great Hall of the People and the Chinese Parliament, close to the southern shore of Zhongnanhai Lake, and from the traditional viewpoint of Jingshan Park, seems to be strangely modernistic and out of keeping with its setting.

FORBIDDEN CITY

Beijing's Forbidden City, or Zijincheng, was established between 1406 and 1420, and remains the best-preserved and most complete collection of imperial architecture in China. It was the home of China's emperors from 1420 to 1911 and, with 9,999 rooms, is the largest such complex in the world. The palace is divided into three sections: the palace gates, the principal halls and the inner court. The predominant colors of the Forbidden City are red and yellow, colors widely associated with China.

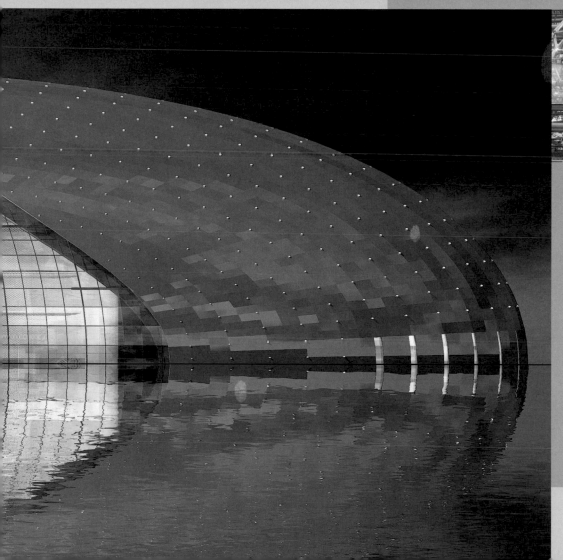

PAUL ANDREU

Andreu (b. 1938) graduated from the École Polytechnique in Paris in 1961, and is best known for his designs of several major international airports, including those at Manila, Jakarta, Abu Dhabi, Cairo, Brunei, and Paris. Other prestigious works by him include the Grande Arche at La Défense in Paris. Andreu's most famous project, the Charles de Gaulle International Airport at Roissy in Paris, was regarded as the jewel in his architectural crown until the collapse of a portion of Terminal 2E in May 2004. Andreu blamed the disaster on poor construction work, but another of his terminals, in Dubai, collapsed in September 2004. He remains a figure of some controversy.

the facts

The final unit of steel hoisted to complete the main arch weighed 441 tons (400 tonnes).

More than 50,700 tons (46,000 tonnes) of steel were required to build the bridge.

Chinese basketball star Yao Ming led a run across the bridge to mark the inauguration.

Lupu Bridge

Shanghai is an important engine for China's continuing economic growth, and nowhere is this more apparent than along the banks of the Huangpu River, which separates the older city of Shanghai from its newer twin, Pudong. For years, under Japanese occupation and Communist austerity, the easiest and fastest way to cross the river was by ferry. No longer. In 1990 Pudong—which means "East of the Huangpu"—was officially designated a Special Development Zone, with Shanghai the "dragon head" of economic development in China. Since then new links have been pushed across or driven under the Huangpu, notably the Yangpu, Xupu, and Nanpu bridges and the Yenan and Fuxing road tunnels.

When the Lupu Bridge opened on 28 June, 2003, it was the grandest of Shanghai's crossings to date and the world's largest steel-arch bridge. At 11,293ft (3,900m) long, it surpassed the previous title holder, the New Virginia Bridge in the United States, by 105ft (32m). The bridge is designed to withstand earthquakes, and allows 150ft (46m) of height clearance and a 1,115ft (340m)-wide navigation lane for river traffic on the Huangpu beneath. Six lanes carry traffic across the bridge from Luban Road in Puxi to Jiyang Road in Pudong, cutting the journey time between Shanghai city center to the new Pudong International Airport from two hours to 30 to 40 minutes.

Recouping the Costs

Shanghai Lupu Bridge Investment Development Company will own and manage the bridge for 25 years. Although no toll is charged to traffic for crossing the bridge, the company expects to recover some of its costs by charging sightseers and selling advertising space on the bridge. The bridge is equipped with four ultra-modern elevators—two in Puxi and two in Pudong—to carry visitors to the top of the great arch for sightseeing. Shanghai local government will also pay the company 9.7 percent of the construction costs annually for the next 25 years.

SOARING SHANGHAI

East of the Bund (Shanghai's waterfront promenade), Pudong rises like an Oriental Manhattan, with some of the tallest and architecturally most innovative buildings in the world. A network of new tunnels, massive suspension bridges, and elevated highways carries the city's ever-increasing traffic across or under the river to Pudong, a huge new extension of Shanghai. Yet Old Shanghai, just north of the Lupu Bridge, has also been restored and is lined with traditional-style dumpling houses, bric-à-brac stores and art galleries. Here, in the shadow of the great skyscrapers, you can stand beneath red Chinese lanterns casting their glow on an intriguing mix of reproduction Chairman Mao memorabilia and other souvenirs.

HUANGPU RIVERS

The Huangpu River is Shanghai's lifeline. Just 61 miles (97km) long, linking the city with the great Yangtze Delta, it provides a major source of drinking water for the thirsty city. Once known to the city's European residents as the Whangpoo, or "Yellow Creek," it is also an important navigation channel, lined with wharves, warehouses, factories, and godowns (warehouses).

87

The world's highest swimming pool, on the 57th floor, helps to counterbalance swaying during earthquakes.

The 5-acre (2ha) site is enclosed within a deep wall made of 27,000 cubic yards (20,500cu m) of concrete.

The spire is designed to resemble a lotus flower, which is a Chinese symbol of good luck.

Jin Mao Tower

Shanghai's newly built business district of Pudong is a place like nowhere else on earth. In the course of less than 20 years a city hinterland of peasant villages and farms has been transformed into a financial center wielding global economic power. From the Oriental Pearl Tower of 1995 to the still-incomplete Shanghai World Trade Center, Pudong's extraordinary architecture has taken on ever more futuristic forms—but the Jin Mao Tower remains its most distinctive building.

Measuring 1,380ft (421m) to the tip of its spire, the tower is one of the tallest buildings in the world and, like other super-skyscrapers in the Far East, stands as a proud statement of modernity and wealth. Although it was designed by Western architects (the Chicago-based firm Skidmore Owings & Merrill), its form is defiantly Chinese. During the dark years of the Cultural Revolution, China's leaders attempted to create a new society with no links to the past. A generation later, in a very different political and economic climate, the Jin Mao Tower suggests a fresh attempt to recreate the splendor of old China in a contemporary form. Completed in 1999, it is based on traditional designs, and yet it resembles a pagoda for the modern age.

LUCKY EIGHTS

In Chinese numerology the number eight represents prosperity and growth, and is considered particularly fortunate in business matters. It is for this reason that the Jin Mao Tower (Jin Mao translates as "shining luxury") is 88 stories high, and that an opening ceremony was held, before the building was complete, on the eighth day of the eighth month of 1988. The lucky number also rules the architectural design, with the tower divided into 16 sections that diminish as they rise by a factor of one-eighth. Aside from any magical significance, this gives a pleasing symmetry and order to the overall design.

A Soaring Atrium

Clad in a richly detailed curtain wall of steel and aluminum rails, the tower appears delicately fragile, as though it were made from timber and bamboo. Yet, given that the engineers could not find bedrock however deep they bored, and that Shanghai suffers from natural phenomena such as earthquakes and typhoons, its construction proved a major challenge. The 1,062 hollow steel piles that support the tower plunge 274ft (84m) underground, and at the time of construction they were the largest ever made.

Above basement level the first 50 floors are open-plan offices with 147,000sq yards (123,000sq m) of floor space, sufficient for 10,000 workers. Then, above two floors of equipment rooms, the building changes character, with the upper stories housing a 555-room luxury hotel. It is here that you find one of the most spectacular indoor spaces in the world—a circular atrium 88ft (27m) wide that spirals up an astonishing 465ft (142m) to the building's roof. The view from the top-floor observation deck down into this vertigo-inducing void is every bit as dramatic as the panorama of the Yangtze River and Shanghai.

CHANGING FORTUNES

With an estimated population of 20 million, Shanghai is China's largest city and the world's busiest port. After the Opium Wars in 1842 it was divided between British, French, and American self-governing communities. Since then its history has been tumultuous. It was seized by Japan in 1937, and Westerners were brutally interned during World War II. Then, when Mao Tse-tung's army took control in 1949, the city was subjected to Communist reforms and lost its dark but glamorous reputation for crime and corruption. In 1967 it was a hotbed of the Cultural Revolution, but has now become a powerhouse of China's economic growth.

the **facts**

Building the Shanghai Maglev required a 48,000-sq yard (40,000sq m) workshop.

More than 1,500 guideway girders were required during construction.

The train windows are designed so that the views of the landscape are undistorted, despite the high speeds.

40 Shanghai's Maglev Train   **Shanghai,** China

Shanghai's Maglev Train

For more than three decades there has been considerable interest in the concept of high-speed transport by means of magnetic levitation, known as Maglev. With this system the vehicle rides on a cushion of air created by electromagnetic reaction between an onboard magnet and another opposing force magnet embedded in a line, or guideway. Initially developed by German and Japanese scientists, this proved to be reliable in a slow-speed shuttle service between Birmingham station and the city's airport in England during 1991.

However, to replace Europe's perfectly efficient existing rail system was unviable on the grounds of both cost and practicality, so the idea was never put to widespread use. This was not the case in China, however, which had a relatively underdeveloped rail network and an expanding economy. In 2001 a contract was signed with the German company Transrapid International to build the 19-mile (30km) Maglev line between Shanghai's Longyang Road and Pudong International Airport.

Faster Than a Bullet
The first official run of the Shanghai Maglev took place on 31 December, 2002, with Chinese Premier Zhu Rongji and German Chancellor Gerhard Schröder on board. The service opened to the public early in 2004 and has cut travel times between the city center and its main airport drastically. The journey takes at least 45 minutes by taxi, but only eight minutes by Maglev train. An elevated, double-track guideway carries Maglev trains at a maximum speed of 270mph (430kph), although the Maglev is capable of traveling at 312mph (500kph), 60 percent faster than the Japanese Bullet Train. At present, the Shanghai Maglev runs three trains an hour, 18 hours a day.

PUDONG NEW CITY

Until the early 1980s the area east of the Huangpu River—Pudong—comprised mainly rice paddy fields and marshland. But since the extraordinary boom resulting from China's entry into a free market economy, all this has changed. Today Pudong rises like an Oriental Manhattan, featuring some of the tallest and most architecturally innovative buildings in the world, as well as some of the most valuable real estate. A network of new tunnels, huge suspension bridges and elevated highways carries the city's traffic across or under the Huangpu River to Pudong Airport.

Expanding the Network

The trains are sleek and futuristic, and riding in them is smooth, quiet and comfortable—not unlike being in a plane skimming along just above the ground. Acceleration to 188mph (300kph) takes just over two minutes.

A proposal to build another Maglev line between Shanghai, China's major port and industrial center, and Beijing, the national capital, was under consideration in 2006. Despite the high cost, the Chinese authorities are keen for China to make a fundamental contribution to the world's future development in this area of magnetic levitation transportation technology.

A New Era

Following the success of the Shanghai Maglev, plans are now underway to build a Maglev line linking Germany's city of Munich to its airport, as well as in the United States between Anaheim, California, and Las Vegas, Nevada. To test the viability of this idea a 25-mile (40km) line will be built between Las Vegas and Primm.

the facts

The Hangzhou Bay Bridge will cost an estimated US$1.42 billion.

When the bridge first opens, traffic flow is expected to reach 52,000 vehicles per day, rising to 96,000 by 2009.

Two thirds of all China's national private investments originate in the Province of Zhejiang.

Hangzhou Bay Bridge

The Hangzhou Bay Bridge—or, more precisely, the Great Trans-oceanic Hangzhou Bay Bridge—is a cable-stayed structure under construction across Hangzhou Bay. On completion in 2008 it will provide the first direct road link between Shanghai and Ningbo, in the Zhejiang Province. It will also be the longest trans-oceanic bridge in the world.

Construction work started on the bridge in June 2003 after more than a decade of feasibility studies. The bridge will be 22 miles (36km) long with six expressway lanes in two directions, making it the second longest bridge in the world after the Lake Pontchartrain Causeway in Louisiana. The bridge is designed to last for 100 years and to carry traffic traveling at a maximum speed of 62mph (100kph).

The highway will run southwest from Shanghai to the city of Jiaxing, in Zhejiang Province, before swinging southeast, directly across Hangzhou Bay, to the city of Cixi, also in Zhejiang Province, then continuing southeast to the major city of Ningbo. When completed, the route will cut the road distance between Shanghai and Ningbo from 250 miles (400km) to 50 miles (80km).

Economic Growth

One important economic consequence of this is that Ningbo, with its up-to-date container port at Beilun, will be able to compete on more equal terms with Shanghai's vast docks at Pudong for international maritime freight. To the north of Hangzhou Bay, at Jiaxing, and to the south, near Cixi, large new industrial zones are being built to add to the already massive capacity of the Yangtze Delta region. Another economic plus is that tourism, especially in the domestic market, is expected to increase between Shanghai and Ningbo, with driving time cut from four hours to one.

SECOND BAY BRIDGE

In 2003 the Zhejiang provincial government approved another trans-oceanic bridge over Hangzhou Bay. This one will not be anywhere near as long as the first, but at 8 miles (13km) it nevertheless represents a major engineering feat. The bridge will save more than 62 miles (100km) on the journey between Shanghai and the newly industrialized city of Shangyu, in Zhejiang Province. It should be completed by 2008.

Refueling Island

Drivers traveling across the bridge will be able to turn off midway to an ultra-modern services island for rest, food, and fuel. The island will be raised on piers so as not to impede the flow of currents in the bay. The construction of the bridge is part of China's continuing investment in its transportation infrastructure, as well as a way of helping Ningbo and northern Zhejiang merge into and benefit from the Greater Shanghai economic area.

SOURCE OF PROSPERITY

China's awe-inspiring growth rate of the last two decades, about 10 percent per annum, owes much to the relatively small area around Shanghai, the Yangtze River Delta. Shanghai's own importance to China's economy and, by extension, the world's economy, has been well documented, but less well known is the importance of the surrounding region. The delta covers a relatively small area of 38,650sq miles (100,00sq km) and is home to more than 10 percent of the country's population. In 2002 it accounted for 22 percent of China's gross domestic product, 25 percent of revenues, and 29 percent of the country's import and export volume.

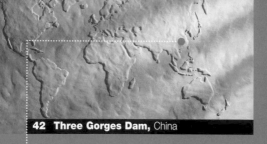

The damming of the mighty Yangtze River in China is one of the most ambitious and controversial engineering projects ever undertaken. When the world's largest hydroelectric scheme is finally completed in 2009 the benefits will be huge, with around 8 percent of China's massive power needs being supplied from a single, clean, renewable resource. Nonetheless, the project has attracted international criticism because of its enormous economic, ecological, and human costs.

Three Gorges Dam

The Yangtze is the world's third-longest river, after the Nile and the Amazon, and more than 400 million people live along its 3,937-mile (6,350km) course. Some 15 million of these people inhabit the floodplain downstream from the picturesque Three Gorges, which cut a spectacular route through 120 miles (200km) of limestone mountains to the east of the city of Chongqing.

Over the past 100 years, floods have resulted in around a million deaths, and a dam to control the river was first proposed by the nationalist leader Sun Yat-sen in 1919. From the 1950s onward Mao Tse-tung endorsed the project, but it was only after Mao's death in 1979 that China's bid to become an economic super-power provided an incentive to realize the colossally expensive plan.

The site finally selected by geologists and engineers lies near the foot of the Three Gorges, where firm, crystalline rock provides an excellent foundation and the river divides around an island. This allows the worksite to be isolated behind a temporary coffer dam. Since 1993, a workforce of 27,000 people has been laboring day and night, 7 days a week.

The dam contains 22 million sq yards (18 million sq m) of concrete and 2 million tons (1.9 million tonnes) of steel.

350 billion cu ft (10 billion cu m) of rock and earth have already been shifted in the course of construction.

Each of the 26 turbines, which were built by Hewitt Siemens and General Electric, can generate 700,000 kilowatts.

China burns 1.6 billion tons (1.4 billion tonnes) of coal a year and is the world's second-largest producer of greenhouse gases, after the United States.

HIDDEN COSTS

The project has attracted fierce criticism both in China and abroad. According to official figures, up to 1.2 million people will have to be resettled as the reservoir expands, but opponents of the scheme put the actual number closer to 2 million, and claim that inadequate compensation has been offered. In the city of Wan Xian alone, 800,000 people will lose their homes, while 115 further towns and 4,500 villages will be destroyed. Some 1,300 ancient sites will also be lost beneath the rising waters, although a number of important antiquities are being moved to higher ground.

Other causes for concern include the possibility of toxic waste leaking from the 1,600 factories and mines that will be submerged, and the risk that silt accumulation may increase the likelihood of flooding. The Chinese government has vigorously countered all such claims and is spending US$4.8 billion on environmental measures. The effect on tourism cannot yet be quantified, for although the landscape of the gorges will inevitably be greatly changed, the dam itself is proving to be a popular attraction.

THREE PHASES OF CONSTRUCTION

In the first phase of the project, which ran from 1993 to 1997, work was concentrated on the construction of a coffer-dam, which was built from stone and earth in waters up to 200ft (60m) deep. Within its watertight confines the world's largest building site took shape, complete with cement works and steel-assembly yards. In 1998 construction of the main dam began, and by June 2003 this was sufficiently advanced for the coffer dam to be demolished and for the raising of the water level to begin.

By the end of the year, the reservoir had risen by 230ft (70m) and the first of the turbines could be brought on line. A five-step ship-lock was also opened, allowing enormous ships to negotiate the change of levels through gigantic lock chambers, each 920ft (280m) long, that extend along a course of 4 miles (6km).

The final phase of this vast project is due to be completed at the end of 2009. When this final phase is eventually at an end, the water level will be raised by a further 130ft (40m), and both of the generating stations will become fully operational.

Massive Power

According to official figures, the budget for the Three Gorges Dam project is US$25 billion, although some estimates claim that the final cost may reach as high a figure as US$75 billion. Whatever the true figure, the undertaking is on a globally record-breaking scale.

The concrete dam is 1 mile (0.6km) wide and 600ft (180m) high. Behind it, a reservoir is growing larger as each new phase is completed. By 2009 this reservoir will extend upriver for 397 miles (640km), covering 423sq miles (1,085sq km). By this time the dam's 26 giant turbines will be generating 18,000 megawatts of power, the equivalent output of 18 average-sized nuclear reactors.

As the flow of the river will be artificially controlled, the risk of flooding will be greatly reduced. Another benefit is that large ships will be able to navigate their way far upstream, reducing transport costs to China's economically relatively deprived interior.

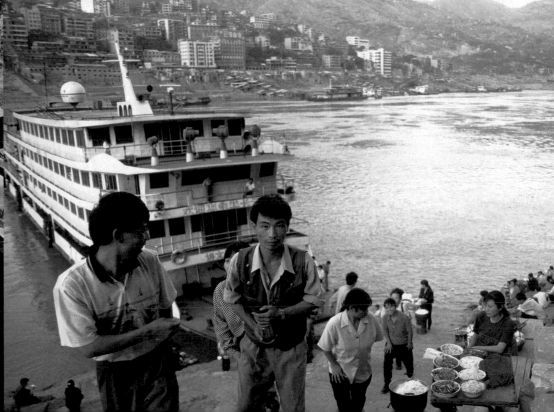

Taipei 101 has 101 stories above ground—hence the name—and five underground.

It cost an estimated US$700 million to build.

A 660-ton (599 tonne) spherical steel damper, installed on the 88th floor, counters earthquakes and typhoons and reduces sway by as much as 30 percent.

Taipei 101

EARTHQUAKE 1999

On 21 September, 1999, central Taiwan was hit by an earthquake measuring 7.6 on the Richter scale. It was centered on the city of Taichung, just 97 miles (155km) south of Taipei, and killed more than 2,000 people. The island sits on one of the most seismically active fault lines in the world, and tremors are an almost daily occurrence down the eastern side of the island. The 1999 quake is supposed to have been of such magnitude that the country's highest peak, Yushan, at 12,963ft (3,952m), is thought to have dropped by 1.6ft (0.5m).

With the booming economy of East Asia there has been constant competition between countries in recent years to build the world's tallest building. In October 2004 Taiwan's turn arrived with the capping off of the massive Taipei 101 tower in the city of Taipei. At 1,667ft (508m), it stands 184ft (56m) higher than the Petronas Twin Towers in Malaysia's capital, Kuala Lumpur, the previous holder of the title, and was designed to put Taiwan on the tall-buildings map. Taipei mayor Ma Ying-jeou said at the opening ceremony: "I have no doubt it can bring Taipei to the world and bring the world to Taipei."

A Giant Pagoda

Taipei 101 is a massive building rising from a comparatively small—169ft (53m) square—"footprint" to provide 240,000sq yards (200,000sq m) of floor space. Resting on giant steel mega-columns designed to resist earthquakes stronger than 7 points on the Richter scale, as well as once-a-century super-typhoons, the four exterior walls of the lower 25 stories slope inward at an angle of about five degrees. Above that, eight stacked, dimensionally identical modules, each eight stories high and with facades leaning out at an angle of seven degrees, rise to level 90.

On top of this 11 mechanical equipment levels step back three times, giving the building an overall appearance of a pagoda, or a Roman candle. A 197ft (60m)-tall pinnacle, rising from the 101st story, completes the tower at 1,630ft (508m). A linked, five-story podium at the base contains a shopping mall. Taipei 101 houses the Taiwan Stock Exchange, and provides office space for 12,000 people. Everything about Taipei 101 is superlative, and the elevators are no exception. Designed by computer giant Toshiba, 34 double-decker shuttle elevators traveling at 37mph (59kph) whisk passengers to the 90th-floor observation deck in under 39 seconds.

OLD AND NEW

In contrast to Taipei 101, the Paoan Temple on Hami Street is Taipei's oldest temple, built in 1765. Usually a riot of color and activity, this Taoist temple was first built by immigrants from Fujien Province. They fled the rigors of life on the Chinese mainland, bringing with them their own deity, Emperor Paoshen, the God of Medicine. Paoshen was, in fact, a doctor from China's Fujien Province, born around AD979, but the local people were so taken with his skills that soon after his death he became known as the "Emperor."

the facts

- The building is designed to withstand 143mph (89kph) typhoons and conforms to seismic-force requirements four times higher than those required in Los Angeles.
- The tower uses only 23lbs (10kg) of steel per square foot.
- The banking hall has a floor area of 1,889sq yards (1,580sq m) and rises 17 stories to an angled glass roof.

Bank of China Tower

British rule in Hong Kong had just seven years left to run when the Beijing-based Bank of China opened a spectacular new regional headquarters in Hong Kong in 1990. Both symbolically and architecturally, it was a building that could not be ignored. Rising 1,209ft (369m) to the top of its twin masts it was, at the time of its completion, the tallest building in Asia and the fourth-tallest in the world. Some viewed its dominating presence as a threat, while others saw it as a reassuring sign of China's commitment to Hong Kong as a gateway to the capitalist world.

Only now, with the 1997 handover an historic fact, can this extraordinary building be appreciated with an open mind. Although no longer Hong Kong's tallest building, it remains an astonishing sight, with its soaring, asymmetric tower of glass with sharply angled walls and roofs that sparkle in the sun like the facets of a jewel. Designed by I. M. Pei in collaboration with his sons, it is a masterpiece of exhilarating geometric forms.

Prisms of Glass and Steel

The tower's most striking feature is the way in which it alters shape as it ascends. At street level, the "footprint" is a perfect square with 170ft (52m) sides. At the 21st-floor level, a quarter of this square is cut away to form a sloping, triangular, glass roof. Further triangles are similarly cut away at the 38th and 51st levels, leaving a single, three-sided prism rising to the 70th floor. The visual effect is superb, with pure, crystalline shapes reflecting the city

SUPERSTITION AND SYMBOLISM

Pei's designs for the Bank of China were inspired by the sectional growth of bamboo, a plant which represents revitalization and hope. Pei's original plans for the building included an X-shaped cross-brace, but he deferred to the traditional belief in Chinese culture that "X" is considered to be a symbol of bad luck and revised his designs.

and the sky. But these shapes also hold the secret of the building's strength. Diagonal steel trusses transfer the full weight of the tower to supporting columns at the corners of the base. A fifth support, braced on legs above the atrium, rises through the upper levels of the tower's core. This innovative system, which was devised by the engineer Leslie Robertson of American engineering firm Leslie E. Robertson Associates, is also highly economical. Despite its extravagant appearance the building cost just US$150 million, which is a fraction of the budget normally required for a project of this scale.

FENG SHUI

Feng shui is a traditional Chinese form of geomancy that dates back 4,000 years. A composite word based on *feng* (wind) and *shui* (water), it represents the invisible (wind, which cannot be seen), and the elusive (water, which cannot be grasped). For millennia the Chinese have used this system to calculate the auspicious siting and alignment of cities, villages, and buildings of all kinds. Although officially illegal in Communist China, geomancy has made a comeback since the liberalization following the Cultural Revolution in the 1960s, and today it is more popular than ever.

At one stage during the construction period there were more than 3,500 people from around the world working on Two IFC.

The 42 high-speed passenger elevators each have a target waiting time of 30 seconds.

The building accommodates 15,000 people.

45 International Finance Center   **Hong Kong,** China

International Finance Center

The International Finance Center, better known as the IFC, is a huge modern office, shopping, and leisure complex on Hong Kong's busy waterfront in the downtown Central District. The first part of the development (completed in 1998) included the One IFC Tower and the South IFC Mall. One IFC is a 39-story, 590ft (180m)-tall skyscraper with 87,200sq yards (72,880sq m) of office space which can accommodate 15,000 workers, while the IFC Mall gathers more than 200 exclusive stores and restaurants, plus a movie theater, under one roof. Neither of these buildings by themselves would make much of an impact in modern Hong Kong; it's the second part of the development, Two IFC, which really sets the pulse racing.

Higher and Higher

When it was completed in 2003, Two IFC, at 1,378ft (420m), was the tallest building in Hong Kong and the fifth-tallest in the world. A gleaming 88-story, glass and steel beauty designed by Cesar Pelli, the skyscraper is home to the Hong Kong Monetary Authority and a number of other world financial institutions. Pelli, renowned for other great buildings in Asia such as Kuala Lumpur's Petronas Towers, has sculpted a shimmering delight with strong, simple lines. The majority of floors have an area of 3,885sq yards (3,249sq m), but as the tower tapers to its apex floor, the dimensions are reduced. In total there are 22 high-ceilinged trading floors, and a sculptural "crown" tops the building. Views from the windows at the top are spectacular, encompassing Victoria Harbor with the Star Ferry chugging backward and forward, and the Kowloon Peninsula and beyond.

But nothing stands still in Hong Kong. In 2007, Union Square Phase 7 will surpass Two IFC as the tallest building in the Special Administrative Zone, and it will eventually stand 1,588ft (484m) high.

FROM BRITAIN TO CHINA

Hong Kong was formally ceded by China to Britain in 1842 under the Treaty of Nanking, and by the end of the 19th century the city had become one of the great ports of the world. The 20th century saw the colony continue to thrive, but toward the end of the century its colonial status had become progressively outdated. In 1984, China and Britain signed a joint declaration allowing for the return of Hong Kong to Chinese rule on 1 July, 1997 under a specially arranged "one country, two systems" strategy. This allowed the zone a high degree of autonomy, including the retention of its own flag and currency.

JARDINE MATHESON BANKERS

The international financial corporation Jardine Matheson & Co, founded by the Scotsmen William Jardine (1784–1843) and James Matheson (1796–1874), is synonymous both with the foundation of colonial Hong Kong and with the fiscal achievements and prosperity of the Hong Kong

Special Administrative Zone. Initially damaged by its association with the disreputable opium trade, and subsequently denounced by the Chinese Communist regime as a symbol of international finance and capitalism, today Jardine Matheson has once again become a respected mainstay of real estate, financial services, shipping, construction, and retail business in emerging, free market China.

103

Nine giant fuel storage tanks hold 50 billion gallons (180 billion liters) of kerosene.

By 2040 the airport is forecast to handle 87 million passengers a year.

Chep Lap Kok means "fish with whiskers", a reference to the red perch that used to be abundant in nearby waters.

Chep Lap Kok Airport

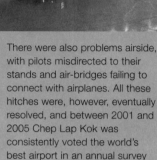

EARLY GLITCHES

The airport's opening in July 1998 was marred by technical disasters on a spectacularly embarrassing scale. More than 10,000 bags missed their flights on the first day alone, information boards failed, moving walkways ground to a halt, and the air conditioning broke down.

There were also problems airside, with pilots misdirected to their stands and air-bridges failing to connect with airplanes. All these hitches were, however, eventually resolved, and between 2001 and 2005 Chep Lap Kok was consistently voted the world's best airport in an annual survey of air-travelers.

Cathedral of Aviation

Yet this was merely the preliminary work to prepare the site. It took a further three years to build one of the largest and most sophisticated airports in the world. The project cost US$20 billion, and during the peak period of construction, 21,000 people were working on the site.

As befits a gateway to the world's fastest-growing economic region, the airport has been planned and built on a truly stupendous scale, with 2-mile (1km) runways that can handle up to 60 flights an hour, and a current capacity of 35 million passengers a year. For the moment there is just one terminal, which makes an astonishing impression even from the air. Designed by Foster and Partners and costing US$1.3 billion, it is the largest covered space in the world, with an area of 564,000sq yards (516,000sq m) and a total length of more than half a mile (1km).

Featuring a roof of undulating vaults that soars 73ft (22m) above ground level, and illuminated by vast walls of glass, the building has been described as a cathedral to aviation. But with up to 100,000 passengers passing through every day, it also has to function as a huge machine, moving people with efficiency and ease. To help the machine run, there are 288 check-in desks, 75 gates, and facilities to handle 19,000 pieces of luggage every hour.

H ong Kong's new international airport began with an astonishing reshaping of the natural world. In 1992, Chep Lap Kok was a small island two hours away from Hong Kong by boat, with steep hills rising to about 300ft (100m), and waves breaking on its lonely shores. A less suitable location for an airport could scarcely be imagined, but in the course of just 31 months the island was totally transformed. More than 44,000 tons (40,000 tonnes) of explosives were employed to blast the mountains into rubble, and the rubble was then used to extend the island out to sea. In all, 258 million cubic yards (197 million cu m) of rock were redistributed and, despite waters 50ft (15m) deep, the island grew at an average rate of 2 acres (1ha) a day to reach four times its former size. By 1995 the island had been turned into a level platform just 20ft (6m) above sea level and measuring 4 miles (6km) by 2 miles (3km).

JUST THE BEGINNING

The airport has brought enormous changes to a district that was previously sparsely populated and remote. A freeway and high-speed trains now run directly from the airport to Hong Kong across one of the world's longest suspension bridges, the Tsing Ma. A new town for 200,000 people is under construction at Tung Chung, on nearby Lantau Island, while within the airport the Sky City complex is due for completion by the end of 2007. Combining a second terminal with hotels, an exhibition center, shopping malls and even a golf-course, this US$2 billion development is part of a 40-year project that will see Chep Lap Kok become busier than London's Heathrow and New York's JFK combined.

On completion of the Metropolitan Buildings, the old City Hall was demolished to make way for the Tokyo International Forum.

2.5 million people visit the observatory every year.

The elevators take 55 seconds to go from the first to the 45th floor in Main Building 1.

Tokyo City Hall

KENZO TANGE

Kenzo Tange (1913–2005) was born in Osaka and graduated from Tokyo University's prestigious Department of Architecture. An admirer of the great Swiss architect Le Corbusier and the Renaissance master Michelangelo, he was also strongly influenced by traditional Japanese architecture. Tenge designed the Peace Park and Peace Center at Hiroshima in 1949, then the Kagawa Prefectural Office in 1958. In the 1960s he designed Tokyo's Cathedral of Saint Mary, and the National Gymnasium Complex for the 1964 Tokyo Olympic Games.

Tokyo City Hall, more generally known as The Metropolitan Buildings, is a massive administrative complex designed by Japanese architect Kenzo Tange and constructed between 1988 and 1991. Consciously planned to provide an eminent landmark representative of the Tokyo Metropolitan Government, as well as a home for the vast bureaucratic machine that employs 13,000 people, its twin towers made it the tallest building in Tokyo when it was completed. Although criticized by some people as overly grandiose, City Hall has become synonymous with Tokyo, and dominates the city's central skyline.

The complex, in the Shinjuku district, consists of three main buildings, known simply as Main Buildings 1 and 2, and the Metropolitan Assembly Building. Main Building 1, which houses the governor's headquarters, is 48 stories high, and Main building 2 is slightly lower with 34 stories. The tops of both towers are set at an angle of 45 degrees, giving the structure a twisted appearance, while the central lower section—the Metropolitan Assembly Building, which is just seven stories high—is set slightly back from the main towers. The concept was to provide a secular cathedral of state, and the complex has been dubbed Kenzo Tange's "Notre-Dame" design.

Shinjuku Skyline

From the top of both towers there are extraordinary views across Tokyo, taking in the four Shinjuku Towers, Shinjuku Station and Central Park, and two separate observation points in Main Building 1 are open to the public. Inside, four service cores stand at the corner of each tower, functioning as vast weight-bearing columns. Opposite the complex, steps lead down to a pleasant, sunken, fan-shaped plaza which has gardens, columns, and statues. Here, another eight-story building, joined to the main complex by two narrow bridges, encircles the plaza. The result is something like an enormous theater, an effect heightened by a huge entrance.

MEIJI ERA

Tokyo became Japan's capital in 1868, at the beginning of the Meiji Emperor's reign (1868–1912). This era marked the end of feudalism and the start of westernization, and the architecture from this time was characterized by a strong influx of Western concepts and techniques, with buildings in stone rather than the traditional wood. Few buildings from this period survive in Tokyo, however, as most were lost through war damage or redevelopment.

The atrium facade consists of 24,000sq yards (20,000sq m) of laminated glass.

The 688ft (210m)-long roof of the atrium is supported by two massive steel columns, 407ft (124m) apart.

The Forum is designed to survive earthquakes in excess of 7.5 on the Richter scale.

International Forum

Seen from the air, modern Tokyo appears vast, mundane, and low-rise. Yet here and there high-rise buildings, some standing alone, some in clusters, pierce the skyline. Moreover, some of these structures are distinctly unusual, examples of the postmodernist architecture that characterized the Japanese bubble economy of the late 1980s. This was a time when the real estate worth of Tokyo was erroneously valued at more than the entire state of California.

The International Forum in Tokyo is an outstanding legacy of this era. Located between Marunouchi, Tokyo's prestigious business district, and the Ginza shopping area, the complex is bounded by two of the capital's busiest stations—Tokyo to the north and Yarakucho to the south. The extensive grounds of the Imperial Palace are a short walk to the west. The main elements of the Forum are a 197ft (60m)-high steel and glass atrium to the west of the site, and a group of block-like buildings which house restaurants, shops, and theaters.

Nautical Design

The most remarkable part of the complex is the giant atrium, which has been likened to a ship from the outside and the belly of a giant whale from within. The elegantly curving glass walls are in marked contrast to the surrounding, rather drab, architecture, while suspended semicircular steel girders hanging from the ceiling are designed to echo the frame of a ship.

Lying between the atrium and the more functional towers to the east is a two-block plaza paved in granite, which serves as the main entrance to the Forum and also provides seating among trees and sculptures. The atrium and other buildings are linked by underground walkways, as well as by several glass-encased catwalks, which are designed to give the visitor a feeling of hanging in mid-air. Under the plaza, a concourse links the Forum to local and regional rail networks.

The Tokyo International Forum is Japan's largest congress center, comprising two theaters, over 7,000sq yards (6,000sq m) of exhibition space, numerous conference rooms, restaurants, shopping malls, and other amenities. The soaring atrium serves as the main reception area, allowing sunlight to enter through the laminated glass walls and ceiling to illuminate the lobby area below. The glass walls are kept sparkling by two cleaning robots with rotating brushes, which are each capable of cleaning 359sq yards (300sq m) a day.

TOKYO'S ECONOMIC BUBBLE

An economic bubble occurs when speculation in commodities causes a price increase, which in turn causes more speculation. Eventually the price of the commodity reaches unsustainable levels, causing the "bubble" to burst and prices to crash. In the late 1980s a property speculation bubble drove land prices in Japan (and especially in Tokyo) to absurd heights. This led to the funding of extravagant architectural projects such as the Tokyo International Forum. The bubble burst on the last day of 1989, signaling the end of two decades of economic growth and a collapse in real estate values.

RAFAEL VIÑOLY

The guiding light behind the Tokyo International Forum, Rafael Viñoly, was born in Uruguay in 1944. From an early age he was destined to become one of the world's great architects, and by the time he was 20 he had already helped found one of the largest design studios in Latin America. Much of his early work took shape in Argentina, but in 1978, troubled by the repressive Argentinean military junta, he relocated to New York. In 1989 Viñoly won the competition to design the TIF, and many experts see this as his major achievement.

The flame is made of stainless steel, weighs 397 tons (360 tonnes), and is 141ft (43m) high.

A covering of simulated gold leaf makes the flame gleam in sunlight.

The flame's shape is one of Starck's signature designs, most often seen as a door handle.

Flamme d'Or

Also known as the Asahi Super Dry Building, this modern marvel is located eastward across the Sumida River from Tokyo's Asakusa Station. By any standards, the Flamme d'Or (Flame of Gold), designed by innovative French architect Philippe Starck and completed in 1989, is an unusual structure. Fashioned like an upsidedown pyramid and ringed with tiny, porthole-like windows that are almost invisible from a distance, the shining black-granite, four-story structure is topped by a huge, golden object. This is intended to represent suds blowing from the top of an iced glass of Asahi Dry beer, as well as the *yakushi* ("remarkable progress") of Asahi Breweries. It is certainly striking.

To the north stands a 22-story tower called the Azumabashi Building, designed by the Japanese company Nikken Seikei. It represents a mug of beer, with slanted white panels at the top suggesting a head of foam.

Business of Beer

Asahi—which means "morning sun" in Japanese—is one of Japan's oldest and most popular breweries, founded as the Osaka Beer Brewing Company in 1889. The Asahi brand was launched in 1892, in the midst of the reformist Meiji era (1868–1912). The company flourished and, despite the travails of World War II, was reborn as Asahi Breweries, Limited in 1949.

In 1958 Asahi introduced the first canned beer to Japan, and in 1982 a collaborative agreement was signed with the German brewery giant Löwenbräu. Yet Asahi's real triumph came in 1987, when the company introduced Asahi Super Dry, an innovation that brought about a revolution in the Japanese beer industry and turned Asahi into brewer-kings.

To make "Dry Beer," nearly all the sugar in the brew is converted into alcohol by prolonging the period of fermentation. The result, which is supposed to have a crisper, cleaner finish and less aftertaste, gained fleeting popularity in the United States through the Budweiser and Miller brands, but has since all but died out. In Japan, however, "Dry" remains the enduringly popular beer of choice, and helped build Asahi's fortunes during the bubble economy of the 1980s.

BREWING BEER

Mankind has made beer for at least six millennia. As early as 4000BC the Sumerians were enjoying up to 16 different varieties, and carvings on Egyptian tombs—still a few thousand years BC—depict with some detail the various stages of the brewing process. And thousands of miles to the east there are written records from 2,000 years ago which refer to the Japanese drink *sake*, produced in a very similar way to beer but using rice, instead of barley or wheat.

PHILIPPE STARCK

Apart from being the driving force behind such large projects as the Royalton Hotel in New York, Starck (b. 1949) is famous for his quirky reinterpretations of everyday household objects. Many of the world's best museums exhibit such items as his citrus juicer, kettle, and toothbrush, plus a whole variety of chairs. As a child Starck spent hours taking apart and putting back together everything he could lay his hands on; years later we are all the richer for his individual and idiosyncratic view of life.

A central computer system in Tokyo carefully monitors and controls all Shinkansen movements across Japan.

Many innovations, including pre-stressed concrete ties, were introduced to cope with the high-speed trains.

In one year, the combined amount of time all Shinkansen trains were late, was 12 seconds.

Bullet Train

EARTHQUAKES AND TYPHOONS

Japan lies in a particularly active earthquake zone, and is subject to regular and severe typhoons. As a result, Japanese construction work is among the most sophisticated and advanced in the world. The Shinkansen is a prime example of Japanese safety technology, withstanding repeated earthquake shocks and typhoons, including the great Kobe earthquake of 1995. Though a train was derailed, it sustained no serious damage and no lives were lost.

The Shinkansen are Japan's world-famous bullet trains, providing a high-speed service between Tokyo and the other main cities on the island of Honshu and Fukuoka, on the island of Kyushu. The first section of the line, a 320-mile (515km) stretch linking Tokyo and Osaka, opened in 1964. This was extended south to Okayama in 1972, and then on to Fukuoka in 1975. Since that time additional lines have been built north of Tokyo to Niigata and Morioka. The network makes extensive use of tunnels, including one that runs beneath the Strait of Shimonoseki, joining the islands of Honshu and Kyushu.

There are plans for the construction of further Shinkansen lines, for example between Nagano and Kanazawa, and between Hakata and Yatsushiro, but the "last frontier" for the Shinkansen remains the northern island of Sapporo, reached by the 33-mile (54km) Seikan Tunnel beneath the turbulent Tsugaru Strait. However, this link is not scheduled for construction until after 2020, and whether such an undertaking will be commercially viable remains to be seen.

Hideo Shima

Hideo Shima (1901–98), the celebrated Japanese engineer who designed and supervised the construction of the Shinkansen, was the son of a railway engineer. Shima graduated from Tokyo University in 1925 and joined the state-run Japanese National Railways. He oversaw the building of the first line as chief engineer until 1963, when he resigned because of escalating costs. He went on to become the president of Japan's National Space Development Agency in 1969, and in the same year became the first Asian to receive the James Watt International Medal from Britain's Institution of Mechanical Engineers.

Times and Speeds

At present around 250 trains a day run on the Shinkansen network, timetabled to depart at 7.5-minute intervals during the morning and evening rush hours. The system is completely electrified, drawing power from overhead cables. The fastest trains, called Hikari (meaning "light"), can make the 665-mile (1,064 km) journey from Tokyo to Fukuoka in around seven hours. Comprising 16 carriages, the Hikari trains are designed to seat 1,000 passengers.

The maximum speed on older sections of track remains 130mph (210kph), while on newer tracks trains reach 160mph (260kph). Today many Shinkansen trains run regularly at speeds of up to 185mph (300kph), records rivaled by the French TGV, the German ICE and now exceeded by the Shanghai Maglev. Desirable speed limits on the Shinkansen may already have been reached, and the emphasis is increasingly on noise reduction, safety, comfort, and efficiency.

Originally intended to carry passenger and freight services 24 hours a day, the Shinkansen now operates passenger services only, and the system shuts down between midnight and 6am so maintenance work can be done. The relatively few trains that run overnight in Japan still use the old narrow-gauge track, which runs parallel to the Shinkansen between major cities.

The triple-level terminal is 470 yards (430m) long by 33 yards (30m) wide.

The building's plate-glass windows are fixed only at their bases, to allow for movement during earthquakes.

225 cruise ships, including the *Aurora* and the *QEII*, berthed at the terminal in 2004.

Yokohama Port Terminal

I n recent years city waterfronts across the world have undergone a glamorous revival, acquiring an appeal that was lacking in the days when they were working docks. Japan's historic port of Yokohama is no exception, with quayside promenades and ultra-modern new developments overlooking Tokyo Bay. But one new feature in particular attracts the eye: a long, low-lying structure with an undulating roof that stretches out into the bay.

The Yokohama Port Terminal is a structure of such astonishing originality that it is not easy to define. It is, as its name implies, the marine equivalent of an international airport and can handle up to four huge cruise liners at a time. But it also functions as a public space. Young couples stroll along its wooden decks, and skate-boarders hurtle along steeply sloping ramps on the roof, which doubles as a park. There is an open-air plaza for markets and summertime events, a multi-purpose indoor hall, plus restaurants, stands, and stores. Bridging the divide between the city and the sea, it is a place where local people intermingle with visitors from every corner of the world.

Origami in Steel

The terminal's bold design derives from its conception as a single structure and the repeated use of three materials: steel, wood, and glass. Steel provides both a skeleton of box girders and a shell of folded plates, allowing column-free internal spans of up to 98ft (30m). Floors and external decks are all made from a Brazilian hardwood that can withstand marine conditions without needing a protective finish. Throughout the building, walls of glass overhung by cantilevered decks blur the line between internal and external space, while sloping ramps, curves, and folds provide a constant sense of fluid movement to the structure as a whole. Internally, steel ceilings have been left unclad,

revealing a geometric pattern of triangular forms reminiscent of the paper-folding art of origami. Working on a larger scale, and with assistance from computer models, the architects have designed a building that unfolds in a sequence of surprises as it is explored from its lower-level parking lot, through cavernous internal halls, to the curious topography of its roof-top park.

Most landmark buildings tower skyward or shout out their presence with architectural display. The Port Terminal adopts a lower-key approach, while still providing Yokohama with an internationally acclaimed masterpiece of engineering and design.

FOREIGN OFFICE ARCHITECTS

The husband-and-wife team of Alejandro Zaero Polo and Farshid Moussavi were little-known young architects when they entered a competition for the Port Terminal's design in 1995. Against all expectations, their entry was judged the winner by a panel of international architects, despite

752 rival proposals. Since then they have represented Britain at the 2002 Venice Biennale and worked on major projects in Britain, Spain, the United States, and Japan. The name they have chosen for their practice is ironic. Born respectively in Madrid and Tehran, they consider themselves to be foreigners in London, and claim that their approach to projects is always a diplomatic one.

YOKOHAMA

The US$200 million terminal was conceived as a high-status civic project that would improve Yokohama's prestige. However, since the end of World War II, Japan's foremost sea port has increasingly been swallowed up by its larger and more powerful neighbor, Tokyo. The project came close to being cancelled during the Far Eastern economic crisis of the late 1990s but was revived, largely in response to Yokohama being chosen as the venue for the World Cup football final of 2002. Since the terminal's opening in April of that year, the city has become one of Japan's major tourist destinations.

Kansai International Airport

MANMADE ISLANDS

Artificial islands such as the one on which KIX is built seem to be the apex of modernity. Yet nothing could be further from the truth. Artificial islands have a long history in many parts of the world, with surviving examples such as the crannogs of prehistoric Scotland and Ireland, the floating islands of Lake Titicaca in Bolivia, and the ceremonial island center of Nan Madol in Micronesia. There are also the countless artificial islands, called chinamitl, that surround the extraordinary city of Tenochtitlan, the predecessor to Mexico City. These man-made islands surrounded a small natural island in Lake Texcoco.

Kansai International Airport (KIX) opened in 1994, and is one of the world's most ambitious construction projects. Designed to serve the Osaka-Kyoto region, it is named after an old term for central Japan—*kansai*, or "west of the barrier"—to differentiate it from the Tokyo region. The complex is situated on an artificial island 3 miles (5km) long by 2 miles (3km) wide, 3 miles (5km) offshore in Osaka Bay.

The double-decked Sky Gate Bridge, with a highway on the upper deck and a railway on the lower, connects the island to the mainland. Being offshore, KIX was able to function as East Asia's first 24-hour hub, linking around 30 central Japanese cities to 40 countries around the world. It handles around 400 international flights a week. KIX took almost eight years to complete at a cost of US$14.4 billion, but expansion of the artificial island is already well underway. This will add 2sq miles (5sq km) of new land, intended to accommodate an additional terminal and two brand new runways.

Making Waves

Dominating the airport complex is the single four-story terminal, designed by well-known Italian architect Renzo Piano (b. 1937). At a mile (1.7km) long, this is the longest building in the world. A giant atrium rises 200ft (60m) from floor to ceiling, with escalators and glass-fronted elevators connecting the four levels. The first level is for international arrivals, the second for domestic flights, the third for storess and restaurants, and the fourth for international departures. Passengers are moved from one end of the terminal to the other by a sophisticated people-mover system.

However, the single most spectacular feature of the terminal is its beautiful, undulating roof, which is made from 82,000 identical panels of stainless steel. Its futuristic shape is designed to promote air circulation through the building, as can be seen by watching the huge mobiles by Japanese sculptor Susumu Shingu (b. 1937) that are suspended in the ticketing hall.

KYOTO

In the 1960s, when the Kansai region was rapidly losing business to Tokyo, planners proposed the construction of Kansai International Airport to improve international communications with the industrial cities of Osaka and Kobe. But KIX also serves as the gateway to Kyoto, considered the cultural heart of Japan. Home to an estimated 1,600 Buddhist temples and 400 Shinto shrines, as well as numerous palaces, gardens, and other fine buildings, It was deliberately spared firebombing during World War II because the Allies recognized its special historical significance.

The Sky Building was conceived in 1988 with four interconnected towers, but practical considerations reduced the number of towers to two.

It is the fifth-tallest building in Osaka, Japan's second city.

The Takimi Koji restaurant mall on the basement floor is a replica of a 1920s Osaka street market.

Umeda Sky Building

Located in Osaka's northern commercial district of Kita (also known as the New Umeda City), the Sky Building is one of the city's most recognizable landmarks. The skyscraper was originally conceived as part of a larger City of the Air complex, to be built with funding from the Japanese bubble economy, but it had to be scaled back for economic reasons. Nevertheless, the Sky Building still represents the ultra-wealthy, high-tech face of Japan.

Designed by Hiroshi Hara and completed in 1993 with finance from Sekisui House Corporation, the 568ft (173m) building consists of two 40-story towers linked at the top by the Floating Garden, and it is billed on its website as "the world's only pair of skyscrapers connected in mid-air." Visitors travel to the 35th floor via a glass elevator attached to the side of the building. From here, escalators in glass tubes rise diagonally up to the Floating Garden, emerging through a huge circular hole. One thousand nozzles spray fine mist into the air outside the viewing platform, creating the effect of a "sea of light blue clouds," though some visitors find this adds to the alarming vertigo induced by the climb.

Hiroshi Hara

Born in 1936 in Kawasaki, Japan, Hiroshi Hara began his architectural studies at the University of Tokyo in 1959, completing his doctorate in 1964. By 1969 he was an associate professor at the Institute of Industrial Science, University of Tokyo. A glittering professional career lay ahead as both an architect and an author on architecture. Perhaps his most famous work, apart from the Umeda Sky Building, is the controversial Kyoto Station, the city's largest commercial development. Opened in 1997 to celebrate Kyoto's 1,200-year anniversary, the station represented a leap into the future for this historic city. A rash of other modernist buildings has followed, to the annoyance of many citizens.

Garden in the Sky

From the rooftop of the Umeda Sky Building, the 360-degree views of the city are superb. The Floating Garden was assembled on the ground and then raised 558ft (170m) into position, using wire ropes, at the rate of 14in (35cm) per minute—a remarkable feat considering it weighs 1,229 tons (1,040 tonnes). The towers, which are equipped with advanced earthquake resistance and vibration control features, house many stores and restaurants, as well as a movie theater. Outside there is a beautiful garden with sculptures, pools, and fountains. There are also two walking courses at ground-level.

OSAKA CASTLE

Built by Toyotomi Hideyoshi in the 16th century, Osaka Castle has played host to some of Japan's bloodiest events. Hideyoshi, famous for helping to unite Japan after an extended period of civil war, intended the impregnable castle to be a symbol of his authority and power. But by 1603, with Hideyoshi long since dead, the Tokugawa clan longed for the destruction of the ruling Toyotomi clan. In 1614 a Tokugawa fighting force of 200,000 attacked the castle; the Toyotomi were outnumbered 2-to-1 but managed to repel the assailants. However, in the following year another massive Tokugawa force attacked the castle, this time defeating its defenders. The Tokugawa, or Edo Period, had begun.

In 2006 the Osaka World Trade Center became the tallest building in western Japan.

The World Trade Center provides trade and economic data for more than 400,000 companies in 81 nations.

Osaka is Japan's second city after Tokyo.

the **facts**

54 Osaka World Trade Center   **Osaka,** Japan

Osaka World Trade Center

The Osaka World Trade Center, also known as Cosmo Tower, dominates Cosmo Square in Osaka, on the Japanese island of Sakishima. The building lies at the heart of the Technoport Osaka Plan, a mega construction project which is being developed on 1,915 acres (775ha) of reclaimed land in Osaka Bay. Its facilities are designed to function as an office for international trade, a convention center, and to promote the development of advanced technology in 21st-century Osaka. The project also aims to provide housing, as well as cultural and commercial facilities. Cosmo Square is also home to INTEX Osaka, the largest international trade fair venue in western Japan, and the Asia Pacific Trade Center, making the square a hub of international commerce.

Elegant Edifice

Clad in a reflective glass curtain wall with white, vertical mullions, the World Trade Center was officially opened on 20 April, 1995. It is an elegant structure rising to a height of 840ft (256m), with three stories underground and 55 above ground level. Floors seven through to 43 are reserved for offices, while the remainder of the building houses assembly halls, conference rooms, restaurants, shopping malls, banks, and a post office. There's a wedding hall on the 49th floor, while the Osaka World Trade Center itself is on the 50th floor.

Cosmo Tower has its own satellite communication system and underground optical cable network, as well as excellent access to the harbor and Osaka International Airport.

Osaka City

A busy, bustling, metropolis second only in importance to Tokyo, Osaka has a long and influential history. In the 6th century visitors from Korea and China used Osaka (then known as Naniwa) as an entry point to Japan. These visitors bought with them a variety of new technologies in construction, engineering, and forging, but most importantly introduced a new religion, Buddhism. By AD645 Buddhism had taken a firm hold on the country. In the 7th century Emperor Kotoku decided to move his capital from Asuka to Osaka. Since then, Osaka has remained the country's most important gateway for foreign culture and international trade.

NIKKEN SEKKEI

Apart from being Japan's largest independent architectural consultancy firm, Nikken Sekkei is also its oldest. Founded in 1900, it is a company that has always prided itself on being in step with the changing face of modern Japan. It has looked to the wider world over the last century, completing more than 14,000 projects in nearly 50 countries. The company employs around 2,000 people and is rightly famous for such monumental undertakings as the Kansai International Airport terminal, and the Islamic Development Bank Headquarters in Jeddah, Saudi Arabia. The firm is also known for smaller-scale design work, such as Tokyo's Sakuradamon Police Box.

Spectacular Views

Visitors can take a non-stop, 80-second ride in a glass elevator to the observatory—an inverted glass pyramid dubbed the "Top of the Bay"—on the 55th floor. On a clear day it is possible to see Kansai International Airport, Awaji Island, and the Rokko Mountains rising behind Osaka. Looking down onto Cosmo Square, there are fine views of the neighboring Asia Pacific Trade Center and a large waterfront park lined with palm trees, which is designed to create a sub-tropical atmosphere.

A total of 213,000 tons (193,200 tonnes) of steel was used in the bridge's superstructure.

The length of wire used in the cables measures 187,500 miles (300,000km).

When they were built in 1998, the 928ft (283m) towers were the tallest of their kind.

Akashi Kaikyo Bridge

Despite the fact that Japan sits on an active earthquake fault line, the Akashi Kaikyo Bridge was the world's longest suspension bridge when it was completed in 1998. The Japanese government first drew up plans for a bridge to cross the Akashi Strait as long ago as 1955, when a dreadful storm sank two ferries and 168 children drowned. After changes to the plans, which originally included a railway line, construction finally began in May 1988.

The six-lane bridge, known in Japan as the Pearl Bridge, connects Honshu Island to Awaji Island. It is part of an important road link between the two major islands of Honshu and Shikoku. With a total length of 4,377 yards (3,911m) and a central span of 2,177 yards (1,991m), it surpasses the previous world-record holder, the Humber Bridge in England, by 635 yards (581m). At the opening of the bridge on 5 April, 1998, 1,500 invited guests walked across it, and the Crown Prince and Princess of Japan officiated at the opening ceremony.

KOBE EARTHQUAKE

Early on the morning of 17 January, 1995, while many of Kobe's citizens were fast asleep, the biggest earthquake to hit Japan since 1923 rocked the city. Known as the Great Hanshin Earthquake, it measured 7.2 on the Richter scale and caused enormous damage, leaving 300,000 people homeless, and destroying more than 100,000 buildings. The epicenter of the quake occurred almost directly below the Akashi Kaikyo Bridge, and such was its power that the tops of the two towers moved almost 3ft (1m) further apart. However, as the bridge was still under construction it was not difficult to alter the design.

Thwarting the Elements

At its mid-point, the sleek, two-towered steel suspension bridge hovers 318ft (97m) above the surface of the Akashi Strait. Prior to construction a 1:100 scale model was built, which underwent a variety of severe test scenarios. The tests proved that the bridge can resist 180mph (288kph) winds and earthquakes registering 8.5 on the Richter scale.

The strait is one of the world's busiest shipping lanes, so, as well as being able to withstand typhoons, earthquakes, and tsunamis, the bridge had to be designed in such a way that it wouldn't impede the river traffic beneath. Due to this and international maritime safety laws, the width of the waterway between the two towers needed to be a clear 1,664 yards (1,500m).

LIGHTING UP

By day, the green-gray bridge harmonizes with its semi-urban environment, and the color works well with the shadows of the sea and sky. But by night the bridge takes on a life of its own, sparkling with 1,737 red, green, and blue lights. Using advanced computer technology, the illuminations form 28 different patterns, which are changed on a monthly basis.

Next door to the Sea Hawk is the impressive Fukuoka Dome, home of the Fukuoka Daiei Hawks baseball team.

The total floor space of the Sea Hawk is 165,000sq yards (138,155sq m).

All the bathrooms are on the building's seaward side, to create the sensation of being in an open air spa.

the **facts**

Sea Hawk Hotel

Perhaps the most spectacular building to be found on Kyushu, the southernmost of Japan's four main islands, is the Seahawk Hotel and Resort at Fukuoka. Built facing west across historic Hakata Bay, the hotel, like Fukuoka City, looks toward the sea, the Korea Strait, and the Asian mainland. The site is redolent with history. It was here that invading Mongol fleets were repelled by typhoons known as *kamikaze*, or "divine wind," which destroyed the ships.

Glass Atrium

Glass features prominently on both the facade and in the overall motif of the Sea Hawk Hotel. Although the main structure is steel-reinforced concrete rising through 38 floors (two below ground and 36 above), the building is clad in glass, with each of the resort's 1,052 rooms enjoying outstanding views across Hakata Bay. Glass is also the defining feature of the Sea Hawk's remarkable 131ft (50m)-high atrium, which dominates the front of the hotel and is the resort's nerve center.

FUKUOKA FUGU

Fukuoka is known for its *fugu* restaurants. *Fugu*, or Japanese blowfish, is famous throughout Japan and beyond as a delicacy, but, if improperly prepared, its flesh can prove fatal to the diner. This is because the fish contains lethal amounts of the poison tetrodotoxin in its internal organs. Although only specially licensed chefs are allowed to prepare and sell the fish to the public, and the consumption of certain organs, including the liver, is prohibited, numbers of people die every year from *fugu* poisoning.

BLOWN AWAY

In 1266 Kublai Khan, Emperor of China, sent emissaries to Japan ordering the country to submit to Mongol rule or face an invasion. The Japanese declined. The Khan bided his time, building a fleet consisting of more than 800 ships, and recruiting 23,000 Mongol, Chinese, and Korean soldiers. On 20 November, 1274, this armada met the Japanese fleet in what has since become known as the Battle of Hakata Bay. The battle raged all day, but by nightfall a huge storm had blown up, forcing the Mongols to retreat. They tried again in 1281, with exactly the same outcome.

No Ordinary Glass

The glass used at Sea Hawk is of no ordinary variety, as it has to withstand the violent typhoons that buffet Fukuoka from time to time, as well as salt damage from ocean spray, and earthquakes that could register 7 and above on the Richter scale. To cope with these extreme conditions, ultrastrong laminated glass has been employed throughout the building. Even if shattered by earthquake or storm, the huge glass panes will stay in place, held together by the laminated interlayer.

The hotel, which is popular for weddings and conventions, was designed to resemble a great ocean liner heading out to sea. This concept is echoed in the interior, which is based on the theme "Cruising the World." Images and designs on each floor evoke impressions of the continents of the world.

The interior of the giant atrium has spectacular views across Hakata Bay, while at the same time encompassing a peaceful environment filled with fountains, trees, gardens, and sculptures. The atrium itself is made of reinforced glass, which muffles wind noise and helps to maintain a comfortable temperature all year round. To be in the Sea Hawk's atrium during fierce, stormy, weather can be an uncanny experience. All about is calm and still, with waterfalls and classical music clearly audible above the quiet conversation, while outside the elements hurl themselves, ineffectually, against the glass walls of the steel-reinforced building.

There are 1,740 window panes in the building, sufficient for more than 200 town houses.

The pilings for the building go 213ft (65m) below ground, the same as the height of a 22-story building.

The area inside the building is 215,000sq yards (179,400sq m), equivalent to about 30 soccer pitches.

Baiyoke Tower II

The Baiyoke Tower II is located at 222 Rajprarop Road in Ratchathewi, in the heart of downtown Bangkok's business district. When it was completed in 1997, the 90-story tower immediately entered the record books as the tallest building in Thailand. It was also the third-tallest hotel in the world, but has since been surpassed by Burj Dubai and the Jin Mao Tower in Shanghai.

One of Bangkok's best-known landmarks and a useful reference point for visitors to the Thai capital, the tower was designed by the Plan Architects Company and constructed by Multiplex, with structural engineering by the Thai company Arun Chaiseri Group. The overall height of the building is 1,014ft (309m), or—in the unlikely imagery of Baiyoke publicity material—"about the height of 182 people standing one on top of the other." In 1999 a radio antenna was added to the top of the building.

Baiyoke II is home to the huge Baiyoke Sky Hotel, which occupies 52 of the tower's 84 floors between levels 22 and 74. The hotel has 673 guest bedrooms, as well as five

GOLDEN MOUNT

Until Bangkok's recent building boom, Phu Khao Thong, or Golden Mount, was the city's highest point. In the early 19th century the land was occupied by a giant *chedi* (tower, or pagoda) for the Wat Saket temple, but the building collapsed on the marshy ground and the remains were made into an artificial hill. During King Mongkut's reign (1851–68) a small *chedi* was built on top of the mount, and in World War II concrete walls were added to its slopes to prevent erosion. Today the hill provides wonderful views of Rattanakosin Island, the old royal section of Bangkok.

restaurants and coffee shops. There are observation decks on the 77th and 84th floors, and a roof-top bar with music on the 83rd floor. The observation deck on the 84th floor is known as the Sky Walk because it features Thailand's only open-air, 360-degree-view revolving roof deck, which can be a windy and invigorating, but also vertigo-inducing, experience. The panoramic views of Bangkok from the deck are unparalleled, and it takes about five minutes for the floor to revolve slowly through its full circle.

A small charge is made for anyone wishing to visit the observation deck without dining in the restaurant, but the views of the city are certainly worth it.

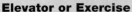

Elevator or Exercise

Getting to the top is fast and simple—two elevators whisk visitors to the top of the building, with a change of elevator at the 77th floor. On the other hand, if you are feeling energetic it is possible to climb to the top via the 2,060 steps, which takes most people an hour or more. Baiyoke II overshadows its predecessor, Baiyoke I, which was similarly named for its developer, Panlert Baiyoke. At just 44 stories, Baiyoke I is best known for its unusual use of multicolored balconies to create a rainbow-like effect.

SPIRIT HOUSES

Practically every building in Thailand has its own spirit house, which provides shelter for the guardian spirits who watch over and protect homes and businesses. They are built outside, and it is important that the shadow of human habitation doesn't fall on them. Offerings such as flowers and food are placed at the spirit house, and incense and candles are lit daily. Not to consult and inform the spirit about business matters is considered to bring extremely bad luck.

Each of the four pylons is supported by 230ft (70m)-deep piles.

Concrete had to be poured for 30 hours nonstop to complete the pylons' foundations.

The project is using the world's largest movable scaffolding system.

Bangkok Mega Bridge

Bangkok, blighted by some of the worst traffic jams in the world, has undertaken several massive new infrastructure projects since the mid-1990s. An elevated train service known as the Skytrain opened in 1999, along with a new underground metro line in 2004. Both were meant to ease the city's burden, but with fewer roads than other cities of a similar size, Bangkok continues to groan with too many cars, buses, and other vehicles. The average speed in central Bangkok during peak hours is now down to 8mph (13kph). With a population of 10 million, the city needs new roads.

The Bangkok Mega Bridge Project is part of the larger Industrial Ring Road (IRR) Project, an idea originally conceived by the Thai King, Bhumibol Adulyadej. In the mid-1990s, the Thai government approved the plan. A new ring road was envisaged which would connect Klong Toey (Bangkok port) with the industrial southern outlying suburbs of the city. The road, including 4 miles (6km) of elevated expressway with eight lanes for traffic, needed to cross the mighty Chao Phraya River, which at this point virtually doubles back on itself, creating the need for two large road bridges.

RIVER OF KINGS

Bangkok is dominated by the Chao Phraya River (the River of Kings), which rises in the north of Thailand as four separate rivers, the Ping, Nam, Yom, and Wang. These come together near Nakhon Sawan, some 156 miles (250km) north of the city, eventually sweeping through the capital on its way to the sea. On its journey it waters the fertile central plains and provides a major natural highway for shipping. Bangkok's extensive network of canals is also fed by this mighty water system, and in times past the city was widely known as the Venice of the East. No more, perhaps, as now the city is ringed in steel and concrete.

Work on the two cable-stayed bridges, with lengths of 768 yards (702m) and 638 yards (582m) respectively, began in March 2003. The bridges are 167ft (51m) high, allowing easy access for all major container traffic plying the Chao Phraya River.

Making Way

Between the two bridges a huge 164ft (50m)-high intersection allows traffic to flow freely to the west or across either bridge in a north–south axis. The area beneath the junction is to be landscaped and redeveloped into a leisure area, where new facilities will include a museum and a 144,000-sq yard (120,000sq m) central park. However, the scheme has not been without human cost. Almost 900 private residences and factories have had to be torn down to make way for the giant bridges, and ferries have had to be rerouted, causing enormous inconvenience to large sections of the southern Bangkok community. The cost of the project, which is due to be completed in 2006, will be in excess of US$220 million.

BAMBOO CITY

When Polish-born British novelist Joseph Conrad (1857–1924) first visited Bangkok more than a century ago, he wrote about the buildings that he saw. He was amazed that an entire city could seemingly be constructed of bamboo and leaves, and said that the buildings were built with not even six pounds' worth of nails. How things have changed.

Today Bangkok is founded on concrete, held rigid by thousands of tons of steel. And yet bamboo still has its place: many of the sledgehammers wielded on building sites have bamboo handles, as they combine strength and flexibility, permitting more forceful blows than usual. Even the scaffolding may be of bamboo, rising precariously for dozens of stories up the latest gleaming tower block.

Both towers have 29 double-decker high-speed elevators, which each carry a total of 26 people.

The superstructures of the towers contain 5,650,000cu ft (160,000cu m) of of concrete.

At the peak of construction just over 7,000 workers were on site at any one time.

Petronas Towers

Skyscrapers are not often seen as objects of classic beauty. There are however, a few notable exceptions, Manhattan's famous art deco Chrysler Building and the John Hancock Building in Boston being just two. Undoubtedly, the Petronas Twin Towers in Kuala Lumpur, completed in 1998, fall into this category. In 1996 the towers, both standing at 1,483ft (452m), were declared the tallest buildings in the world, a position they held until 2003 when they were surpassed by Taipei 101 in Taiwan.

CESAR PELLI

One of the most highly respected architects working in the world today, Pelli, born in Tucumán, Argentina, first studied architecture at the Universidad de Tucumán. After moving to the United States in 1952, he finished his studies at the University of Illinois. His early career saw him involved in a number of large airport projects, most notably the John F. Kennedy Airport in New York City. During the 1970s his work moved away from the traditional load-bearing masonry style and he began perfecting the technology of glass skins, almost like veiling a building. Pelli's World Financial Center and Winter Garden in New York are fine examples of this style.

The Selangor Turf Club, established in 1896, originally occupied the area on which the towers now stand, but due to exceptionally heavy weekend traffic, caused mainly by visitors to the racetrack, the government ordered the club to relocate outside the city. The resulting piece of prime real estate attracted many different plans for redevelopment. The original brief sent out to eight architectural firms chosen to compete for the prestigious assignment, stated the need for "a building that would be identifiably Malaysian, that was of world-class standards and which Malaysians could be proud of".

Islamic Design Elements

The competition was won by the Argentinean-born American architect Cesar Pelli, who designed the twin towers to be identical in all respects. In a statement of Malaysia's primarily Muslim heritage, the 88-floor stainless steel and glass structures were intended to reflect designs found in traditional Islamic architecture. The buildings' exteriors are clad in 700,000sq ft (65,000sq m) of stainless steel and 830,000sq ft (77,000sq m) of glass, and an enormous amount of concrete was used in the construction. The foundation of each tower contains 466,000cu ft (13,200cu m) of reinforced concrete. Each tower is supported by 16 columns made of steel-reinforced concrete around their perimeters and topped with a pyramid-shaped crown.

At the 41st and 42nd floors, some 558ft (170m) above street level, a steel skybridge, 63.6 yards (58.4m) long, connects the two towers. Trips to the skybridge are free to the general public, but they are limited to a certain number of people each day. The bridge played a starring role in the climax of the Hollywood movie *Entrapment* (1999), featuring a cat burglar played by Sean Connery opposite Catherine Zeta Jones. Today Tower One houses Malaysia's national petroleum company, Petronas, and Tower Two is leased out to multinational companies including Microsoft. In 2004 the towers won the prestigious Aga Khan Award, a prize awarded annually for outstanding architectural design in the Muslim world.

SPIDERMAN

By 1997, Alain Robert had gained an unparalleled reputation for urban climbing. Others have climbed large buildings, but always with the aid of some sort of climbing equipment. Alain Robert shuns any use of such equipment and climbs with only his bare hands. Malaysian authorities had expected the Frenchman to make an attempt on the towers, but were still taken by surprise when on the morning of March 20, 1997, the small figure of Robert was spotted several stories up the building. He usually makes it to the top before being arrested and fined, but on this occasion he only made it to the 60th floor before being arrested.

- Former Malaysian Premier Datuk Seri Dr Mahathir Mohamad opened the building on 10 February, 2003.

- The building can house 7,000 occupants and has 22 sky gardens.

- Menara Telekom cost an estimated $160 million to complete.

THE GENTING HIGHLANDS

Visible from the Menara Telekom on a clear day is Kuala Lumpur's very own Las Vegas, the Genting Highlands. This mystical gamblers' retreat, usually referred to by locals as simply Gentings, is Malaysia's only legal gambling house. The idea for the building was conceived by Malaysian businessman Tan Sri Lim Goh Tong in the 1960s, and the Disney-like complex took seven years to complete. At the time many believed cutting a road through the dense jungle to the 6,560ft (2,000m) highlands was an act of folly in itself, but the complex, which includes a number of hotels and theme parks, has proven extremely lucrative and become very popular with the citizens of Kuala Lumpur.

Menara Telekom

Completed in 2001, just three years after the ground-breaking Petronas Twin Towers were built, the appearance of the Menara Telekom building added yet new luster to the futuristic skyline of Kuala Lumpur (KL), the Malaysian capital. The tower was designed as the new corporate HQ of Telekom Malaysia Berhad, Malaysia's leading telecommunications company. On completion, the 1,107ft (310m) high building of 77-storys on Jalan Pantai Bahru was the city's third-highest structure, and the 20th-tallest building in the world.

Based on the award-winning masterpiece *Pucuk Rebung* (Bamboo Shoot) created by Malaysian sculptor and artist Latiff Mohidin (b. 1941), the Menara Telekom employs curvilinear architecture to represent a young bamboo shoot, strongly founded at the base, but with young leaves sprouting. This is a familiar image in Malaysia, where bamboo shoots are a popular and traditional food.

An Intelligent Building

As befits a telecommunications tower, Menara Telekom is classified as a six-star "intelligent building" by KL City Hall, providing infrastructure for multimedia services with high-speed connectivity, and an energy-efficient facilities management system. The tower has an Integrated Building Management System (IBMS) created by Telekom Malaysia's research and development division to provide an efficient, productive, and cost-effective environment. The IBMS can integrate 11 key mechanical and engineering sub-systems within the building, including air-conditioning, lighting, ventilation, security, and elevator systems. There is also a highly advanced document conveyor system, which transports documents anywhere within the building.

Despite its size, Menara Telekom has been designed to provide a comfortable working environment, with numerous internal landscaped terraces providing quiet retreats for rest and relaxation. In addition to offices, the tower has a theater with seating for 2,500 people, exhibition halls, and sports facilities, including the Menara Telekom Sports Complex (with courts for badminton, basketball, squash, volleyball, and the local sport *sepak takraw*), as well as a large indoor gym.

MUDDY RIVER MOUTH

Kuala Lumpur, which means "muddy river mouth" in Malay, was founded in 1857 at the confluence of the Gombak and Kelang rivers. The settlement began when a tin mine was established in the Klang Valley. As the settlement grew in importance the British rulers of Malaya appointed a head man, Kapitan Cina, to administer it and ensure law and order. By 1880 KL was the capital of Selangor State, and in 1963 it became the capital of the Federation of Malaysia. Today skyscrapers dominate the skyline and the city–formerly a rather languid colonial outpost–has become one of Southeast Asia's most vibrant cities.

- Imported from Sweden, the roof consists of 1,056,056 self-cleaning, glazed off-white tiles.
- The concert hall organ is the largest mechanical-action organ in the world, with more than 10,000 pipes.
- HM Queen Elizabeth II formally opened the Opera House on 20 October, 1973.

Sydney Opera House

The Sydney Opera House is one of Australia's most recognizable and iconic national symbols. Next to the impressive Sydney Harbor Bridge, the opera house appears like a beautiful set of sails gliding over the harbor.

Sir Eugène Goossens, resident conductor of the Sydney Symphony Orchestra, first proposed the idea of building a major new arts venue in Sydney in 1947, an idea that then gained the support of the premier of New South Wales, the late J. J. Cahill. A committee was formed to take the idea forward, and in 1955 it launched a competition to find a design for the project, which attracted 233 entries. The Danish architect Jørn Utzon won, Bennelong Point was chosen as the site, and construction work began in 1959. Problems of design, over-running of budgets, and even hints of corruption dogged its progress, but the Opera House was finally completed in 1973 at a cost of $77.3 million, nearly 15 times the original estimate.

BENNELONG

Bennelong Point, the site of the Sydney Opera House, is named after Woollarawarre Bennelong (1764–1813) of the Wangal people, one of Australia's best-known Aboriginals. In 1789 the British captured Bennelong as part of a plan by the then governor, Arthur Phillip, to learn the customs and language of the local Aboriginal people. He and Bennelong became friends and the Aborigine adopted European-style dress and learnt English. In 1790 Bennelong asked Phillip to build him a house on what is now called Bennelong Point. A year later Bennelong provided musical entertainment for the governor, which was the first known performance at this location.

A Showcase of Talent

Regarded as the busiest performing arts center in the world, the Sydney Opera House hosts about 3,000 events a year covering ballet, cinema, musicals, rock concerts, jazz, symphony concerts, opera, and drama. Two million people patronize the Opera House annually. The building houses five auditoriums, five rehearsal studios, 60 dressing rooms, and several restaurants and bars. It is the home of the Sydney Symphony Orchestra, the Sydney Theater Company, and Opera Australia.

The largest venues are the high, vaulted Concert Hall (renowned for its acoustics), which seats 2,679 people, and the 1,547-seat Opera Theater, with an orchestra pit large enough to accommodate 70 musicians. Both feature white birch and brushbox timber. Smaller in scale are the Drama Theater (seating 544), the Playhouse (seating 398), which is also used as a cinema, and the Studio Theater (seating 364), hosting smaller, contemporary arts productions. The former Reception Hall has been renamed the Utzon Room, and there is a new Exhibition Hall. Outside is an open-air space, with the steps up to the entrance of the opera house providing amphitheater-like seating.

JØRN UTZON

Danish architect Jørn Utzon
(b. 1918) attended the Copenhagen
School of Architecture between
1937 and 1942. He furthered his
studies in Sweden and the United
States before working in the office
of famous Finnish architect Alvar
Aalto. In 1956, Utzon won an
anonymous competition for the
design of a performing arts complex
in Sydney. The Opera House was a
bold concept, and well ahead of its
time. Originally conceived to take
just four years to complete, the
project was plagued by delays and
in 1966 Utzon resigned, never to
return to Australia to see his finished
masterpiece. In 2004 he won the
prestigious Pritzker Architecture
Prize, finally receiving the acclaim
he so richly deserved.

2,416 clocks situated all around the building remind members of special events such as Prime Minister's Question Time.

The building has 4,500 rooms.

1.1 million Australian and overseas visitors pass through the Parliament buildings annually.

New Parliament House

New Parliament House, the home of the Australian Parliament and seat of government, was completed in 1988 to coincide with the bicentenary of European settlement in Australia. It sits on Capital Hill in the center of Canberra, Australia's capital, marked by the huge, four-legged, stainless-steel flagpole that has become a symbol of the city.

In 1980 the Malcolm Fraser government initiated a design competition for a new Parliament House. Out of 329 entries, the Italian-American architect Romaldo Giurgola (b. 1920) won. He subsequently moved to Australia, and on Australia Day 2000 he and his family became naturalized Australian citizens. Work began in 1981, and New Parliament House was finally opened by Queen Elizabeth II on 9 May, 1988. It came in over budget, making it the most expensive building in Australian history at the time.

The New Parliament House is situated just below the Old Parliament Building, and although very different in concept and design, their facades are similar, giving them some sort of unity. Much of the building is buried underground, but meeting chambers and rooms for parliamentarians form, when viewed from the air, a huge outer circle enclosing what appear to be two boomerangs. More than 57 acres (23ha) of landscaped lawns and gardens planted with native trees surround the buildings.

OLD PARLIAMENT HOUSE

Soon after World War I it was decided that Canberra would become the new seat of Australian government. A "temporary" Parliament House was proposed, which was planned to last 50 years until a new, larger, House could be built. In the end the old House lasted for 61 years, with the new capital city of Canberra growing up around it. A handsome building in its own right, Old Parliament House rapidly became too small to cater to the workings of government, with numbers of employees rising from 300 to 4,000.

Works of Art

At the entrance to the Parliament Building, the foyer leads into the Great Hall where an enormous 66ft (20m) tapestry, one of the largest in the world, is displayed. Based on a painting by Australian artist Arthur Boyd (b. 1920–99), the tapestry is one of several Australian works of art incorporated into the fabric of the building. Another original piece of artwork is Aboriginal artist Michael Nelson Jagamarra's large mosaic *Possum and Wallaby Dreaming*. This forms part of the Parliament House Art Collection, which includes more than 5,000 works of art and other objects associated with Australian heritage.

In the Members' Hall is a copy of the historic document, the Magna Carta. This is an English charter of 1215 that limited the power of English monarchs, and the copy was purchased by the Australian government in 1952.

MICHAEL NELSON JAGAMARRA

In 1998 the Aborigine artist Michael Nelson Jagamarra completed an impressively large mosaic, entitled *Possum and Wallaby Dreaming*, in the forecourt of New Parliament House. Jagamarra was born in about 1949 near Pikilyi (Vaughan Springs) and grew up in the bush. As a boy, his grandfather taught him sand and body-painting as well as shield painting, and he began painting regularly in 1981. Most of his works relate directly to Aboriginal Dreaming paths, with enigmatic subjects as *Two Kangaroos*, *Flying Ant*, *Yam Dreaming* and *Lightning Rain*. In 1987 a 27ft (8m) Jagamarra painting was placed in the entrance hall of the Sydney Opera House.

the facts

Snowy Mountains Hydroelectric Scheme

The Snowy Mountains Hydro-electric Scheme was by far the largest engineering project ever undertaken in Australia when it began in 1949. Now, it supplies water to the farming industries of Victoria and New South Wales and about 10 percent of all electricity needs for New South Wales, including Sydney. The objective was to offset the effect of drought in the Murrumbidgee and Murray valleys, by redirecting water to them from the Snowy River and the eastern slopes of the Australian Alps in eastern Victoria and southern New South Wales. Transported through a system of pipes, tunnels, and aqueducts, the water was then to be stored in a series of dams for use in the generation of electricity, and for irrigation.

The resulting system was one of the most complex integrated water and hydro-electric schemes ever built. It links seven power stations and 16 major dams through 91 miles (145km) of tunnels and 50 miles (80km) of aqueducts. Defined as the "largest renewable energy generator" on the Australian mainland, it generates approximately 3.5 percent of the national electricity supply. It also supplies approximately 29 billion gallons (2,100 gigaliters) of water a year to the fertile Murrumbidgee-Murray Basins, providing additional water for the industry representing about 40 percent of Australia's national agricultural production.

The Workforce

More than 100,000 people from more than 30 countries worked on the scheme during construction, with up to 7,300 workers at any one time. About 70 percent of all workers were immigrants, attracted to Australia by new opportunities and higher wages shortly after World War II had devastated Europe, although work in these conditions was dirty, wet, noisy, and dangerous. Whole towns were established to accommodate workers, who were sometimes joined by their families. After the completion of the project most of these workers migrated to the big cities, and today almost nothing is left of these temporary townships.

BANJO PATERSON

A poet and balladeer, Paterson (1864–1941) wrote mainly about life in Australia's great Outback, although like many city men (after being educated in Sydney he became a solicitor), he held a rather idealized image of rural life. His most famous poem was *Waltzing Matilda*, later set to music and now regarded as the country's unofficial national anthem. Another well-known poem, *The Man from Snowy River*, has inspired two movies, a television serial, and a musical. A scene from the poem appears on the Australian $10 note, along with Banjo Paterson himself.

SNOWY RIVER WILDLIFE

Located in eastern Victoria along the Snowy River, the Snowy River National Park is a remote wilderness of magnificent river scenery, spectacular gorges, and dense forests of native trees. These trees include snow gum woodlands, stands of ash forests, alpine ash, manna gum forests, and rainshadow woodlands.

Among the 250 or so native animal species that thrive in these diverse habitats, 29 are considered rare or threatened. Among these are the brush-tailed rock-wallaby, the long-footed potoroo, the spot-tailed quoll, and the giant burrowing frog.

The site area of the mall covers 31,180sq yards (26,067sq m).

The huge mall stands directly above Melbourne Central underground station for ease of access.

Melbourne Central Online keeps customers up-to-date on entertainment, special promotions, and events.

Melbourne Central

THE HODDLE GRID

Melbourne's central business district is based on a grid layout created by surveyor and artist Robert Hoddle (1794–1881). The grid is one mile (1.6km) long by half a mile (0.6km) wide, and aligns with the Yarra River. Hoddle first came to Australia in 1823, spending much of his time surveying the Blue Mountains in New South Wales. In 1837 he was awarded the post of senior surveyor in Port Philip, from where he undertook the layout and design of Melbourne, Geelong, and Williamstown. At the time his city and town plans were considered to be groundbreaking.

M elbourne Central is promoted as the "Colosseum of the Consumer," in a city that likes to see itself as the most sophisticated metropolis in the southern hemisphere. The complex of buildings, designed by Kisho Kurokawa and constructed between 1986 and 1991, is located in the historic heart of downtown Melbourne's business district. It comprises a high-rise office block and a futuristic shopping mall, plus offices, and entertainment complexes.

The Melbourne Central Office Tower, which stands 53 stories high, looms over the nearby shopping mall. Its facades, composed of a variety of materials including aluminum, stone, reflective glass, and tinted glass, have a distinctive mullion pattern.

Japanese influence at Melbourne Central is strong. This is not just because the building's architect, Kisho Kurosawa, and the giant Daimaru store are both Japanese, but also because Japanese developer Kumagi Gumi financed the entire venture.

Glass Cone

Enclosing the shopping complex, which has more than 160 specialty stores as well as the large Japanese Daimaru Department Store, is an enormous 20-story glass cone. Inside, the brick-built Coop's Shot Tower, an Australian Heritage building dating from 1894, has been preserved. It is the only remaining feature from the 19th-century Lead Pipe and Shot Factory which used to stand here. Also inside the cone, replicas of a hot-air balloon and a copy of the Wright brothers' vintage bi-plane, both piloted by mannequins wearing period clothing, are suspended above the shoppers.

Nearby, a gigantic fob-watch by Seiko chimes out the tune of Australia's unofficial national anthem, *Waltzing Matilda*, every hour on the hour, while mechanical cockatoos, galahs, and two cherubs appear and move in time to the music. At the back of the watch is a clear panel where you can see toy koala bears amongst the workings of the giant watch.

KISHO KUROKAWA

Kisho Kurokawa (b. 1934), Melbourne Central's architect, was born in Nagoya, Japan and studied at Kyoto University before moving on to Tokyo University School of Architecture. After studying under Kenzo Tange, famous for Tokyo's City Hall and the Peace Memorial Park in Hiroshima, Kurokawa helped found the Metabolist Movement in 1959. This was a radical avant-garde group of like-minded architects and city planners. The group's ideas were characterized by large-scale, adaptable, and extendable constructions, built for a mass society. Kurokawa has been involved in numerous large-scale projects in Japan and overseas, including the Kuala Lumpur International Airport in Malaysia, and a new wing for the Van Gogh Museum in Amsterdam, in the Netherlands.

Africa and the Mid-East

Although not traditionally known for innovation in the fields of architecture and engineering, in recent years this region has caught up with the rest of the world in spectacular style. Until fairly recently, Egypt's Aswan High Dam, which was completed in 1970, was one of the largest engineering projects to have been undertaken on the African continent. The leader of the recent phenomenal building boom, however, is Dubai in the United Arab Emirates. The realization that Dubai's oil reserves will be used up in just a few years' time, has led to the authorities creating a high-end tourist industry from little more than desert and scorched coastline. Visitors are now flocking to the emirate, attracted by vast, decadent hotels filled to their skyscraping-roofs with sumptuous suites, shopping, and spas.

But this region is not solely devoted to luxury hotels and shopping malls. At the far western edge of the traditional Islamic world is one of the largest and most beautiful mosques in the world, Casablanca's Mosque of Hassan II. It was designed to appear as if floating on water, can accommodate up to 25,000 worshippers, and is like a modern wonder of the old world. Similarly, a new university, complete with mosque and landscaped gardens was built in an undeveloped area at the foot of the arid Oman Mountains in the mid-East, bringing opportunity where previously there had been none.

Looking to the future, Dubai is set to break the record for world's tallest structure when its latest super-skyscraper, Burj Dubai, is completed. The developers have not revealed the exact height of the building, so that any contenders for the record cannot build a structure which will supercede it.

At 688ft (210m) the square-shaped minaret is the tallest in the world.

25,000 worshippers can pray within the mosque, with room for a further 80,000 outside in the courtyard.

More than 2,500 of Morocco's finest master artisans were engaged in the construction work.

Hassan II Mosque

When Casablanca's monumental Mosque of Hassan II was inaugurated on August 30, 1993 it was the world's second-largest mosque (the largest being the Masjid al-Haram in Mecca), and the only mosque in Morocco open to non-Muslims. Work had begun on the mosque–originally planned to commemorate King Hassan's 60th birthday–in 1980. The King's dream was to build one of the world's greatest mosques at the western edge of the traditional Muslim world. He wanted it to be a symbol of the hopes and aspirations of a newly emerging nation and its people. The architect chosen to realize this vision was a non-Muslim Frenchman, Michel Pinseau.

Inspiration for the mosque came from a verse in the Qu'ran, the Muslim holy book: "and His throne was on the water." Designed to appear as if floating on water, the mosque occupies 54 acres (22ha) of land reclaimed from the Atlantic Ocean on the outskirts of Casablanca's most densely populated area.

National Asset

The huge main prayer hall is so large that it would accommodate all of St Peter's Basilica in Rome. Enclosed most of the year, the hall has a sliding steel roof 200ft (60m) high, weighing 1,100 tons (998 tonnes), the underside of which is decorated with cedar wood from Morocco's Atlas Mountains. On hot summer days this roof can be retracted in just over three minutes. Hanging from the roof are numerous chandeliers, seven of which weigh 2,640lb (1,200kg) each. Covered mostly in marble, the floor of the main prayer hall is heated in wintertime. Reinforced glass sections allow worshippers to gaze down on the sea below. The mosque houses a *madrassa* (religious school), conference rooms, a museum, and a library.

MICHEL PINSEAU

After graduating in 1956 from the National School of Fine Arts in Paris, Pinseau (1924–99) designed a number of spectacular buildings. Among these were apartments and offices, notably on the Champs-Elysées in Paris and at the French ski resort of Tignes. In the 1970s he met King Hassan II and this turned out to be a pivotal point in his professional career. For more than 20 years Pinseau was the personal architect of the King, designing the Royal Palace at Agadir, the University of Ifran, numerous administrative buildings in Casablanca, housing estates at Rabat, and even the Moroccan Pavilion for the Seville Exposition in 1992.

MIXED RECEPTION

There was some opposition when proposals were first made for a mosque here, as many believed the enormous expenditure could be put to better use. The project's total cost has never been revealed, but unofficial estimates range between $800 and $1,000 million.

The site chosen for the mosque was a particularly impoverished slum area that might have benefited from an investment of this size. Yet today, few doubt the beauty and majesty of the mosque and are proud of this wonder of the Islamic world.

The vast minaret, which is decorated with exquisite arabesques, is visible from all parts of the city. Almost all of the materials used in the mosque originate from Morocco, the only exceptions being the chandeliers and the white marble used to decorate the area around the *mihrab* (a niche or slab indicating the direction of Mecca). Murano, near Venice, provided the chandeliers, and the Carrara marble also came from Italy. The dominant tones of the building are ocher and green (the colors most often associated with Islam), and two lights, with a range of 22 miles (32km), sit atop the minaret, shining in the direction of Mecca.

The library has enough shelf space to hold at least 8 million books.

Hieroglyphs and characters from 120 different languages decorate the Aswan wall.

The total floor area of the library covers 101,630sq yards (85,405sq m).

the **facts**

Alexandria New Library

HIEROGLYPHS

Hieroglyphs, more commonly known as hieroglyphics, are pictorial characters that represent either the object itself or an idea associated with an object (ideograms), or sound signs (phonograms), which were used for their phonetic value. Inscriptions could be written vertically or horizontally.

Hieroglyphs are usually associated with Ancient Egypt, but they have also been found in other ancient cultures, most notably the Hittite, Mayan, and early Cretan civilizations. Egyptian hieroglyphs were originally used to keep records of a pharaoh's material goods, but gradually became more sophisticated and were used to convey increasingly complex ideas.

This was the world's first globally sponsored library, intended to house the entire written knowledge of mankind. It was a grand, perhaps unachievable, dream, but nevertheless an honorable mission. It seems fitting, therefore, that the New Library in Alexandria should be built near the site of what was once the most celebrated library of classical antiquity.

The original idea for reviving the library came from the University of Alexandria in 1974. The project remained simply an idea until the Egyptian president Hosni Mubarak and UNESCO gave their support. Mubarak proposed a design competition with a $60,000 first prize. In September 1987, after receiving more than 1,400 entries, the sponsors awarded the prize to the Norwegian design firm Snøhetta. Generous financial donations very quickly followed, with many Arab states leading the way, and construction finally began in January 1995. By the time the complex opened in October 2002, the project had cost $220 million, $100 million of which came from foreign donations and the remainder from the Egyptian government.

State-of-the-Art

Built just 131ft (40m) from the Mediterranean Sea, close to the university complex, the library is an 11-story disc, tilted toward the sea and partly submerged in a pool of water, which helps counteract the high humidity found in northern Egypt. Protecting the cylinder is an Aswan granite wall, which, at 525ft (160m) in diameter, was the largest circular diaphragm in the world when the building was completed. To support the enormous weight of the building, 600 bell-bottom piles were sunk to a depth of 131ft (40m) below ground. As the library contains such a valuable collection, the consequences of a fire would be devastating, so it is fitted with the most up-to-date fire-prevention equipment.

The main reading room can accommodate up to 1,700 people over eight terraces, and the tilting roof allows indirect sunlight to flood the building, as well as giving a wonderful view of the harbor and sea.

OLD LIBRARY OF ALEXANDRIA

The most famous library of all time, the Royal Library of Alexandria, was established around the beginning of the third century BC, during the reign of Ptolemy II. Demetrius Phalereus from Athens, a student of Aristotle, became the first curator, and at its height the library was home to around 700,000 manuscripts and scrolls. The library survived for many centuries, but its destruction, believed to be around the third century AD, has always raised controversy. Some believe it was destroyed during Roman Emperor Aurelian's reign, others that it was accidentally ruined during Julius Caesar's invasion in 47–48BC.

The rock-fill in the dam is 17 times greater than the amount of stone used in the Great Pyramid at Giza.

Lake Nasser was named after Gamal Abdel Nasser, President of Egypt from 1954 to 1970.

About 95 percent of Egypt's population of 75 million lives within 12 miles (19km) of the River Nile.

Aswan High Dam

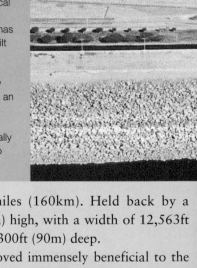

The Aswan High Dam, known as As-Sadd al-Ali in Arabic, is a massive dam that blocks the River Nile just south of the town of Aswan in Upper Egypt, close to the Sudanese frontier. Construction of the dam began, with aid from the former Soviet Union, in 1962 and was completed in 1970. The total cost has been estimated at around $1 billion. Egypt has always depended on the waters of the Nile, to the extent that the country is often called "the Gift of the Nile." This is apparent from the air, where a narrow, fertile, valley can be seen running from south to north across the desert, supporting a population of tens of millions along both banks and throughout the Nile Delta, where the river flows into the Mediterranean Sea.

A Grand Undertaking

A previous, smaller, dam was built across the Nile at Aswan in 1889 under the British administration, and was raised in 1912 and then in 1933, but proved insufficient to control the waters of the longest river in the world. The solution was the building of the new High Dam, about 4 miles (6km) upstream from the old dam. This was a huge undertaking involving the relocation of about 100,000 Nubian and Egyptian farmers. It also entailed flooding the ancient land of Nubia, which spans the modern border between Egypt and Sudan, an area which is extraordinarily rich in ancient artifacts.

With the completion of the High Dam, a new reservoir, Lake Nasser, slowly filled until it stretched back into the

ENVIRONMENTAL IMPACT

The Aswan High Dam has benefited both Egypt and, to a lesser extent, the Sudan by limiting flooding, providing water for irrigating new land, and providing a relatively clean source of electrical power, all of which are good for the environment. Less positive has been the accumulation of rich silt in Lake Nasser, which would normally flow downstream to replenish the already immensely fertile Nile Delta. This has led to an increase in the use of artificial fertilizers and some erosion of the delta, which would normally continue its slow expansion into the Mediterranean Sea.

Sudan for almost 100 miles (160km). Held back by a rock-fill dam 364ft (111m) high, with a width of 12,563ft (3,830m), Lake Nasser is 300ft (90m) deep.

The High Dam has proved immensely beneficial to the economy of Egypt, enabling man to control the annual flooding of the Nile for the first time in history. Hundreds of thousands of acres of formerly arid land have now been brought under cultivation, while the dam's 12 turbines are capable of generating 10 billion kilowatt-hours of electricity annually. Lake Nasser also supports a new and productive fishing industry.

RELOCATION OF ABU SIMBEL

The creation of Lake Nasser meant the submergence of an astonishing array of historical sites from Egyptian and Nubian civilizations. The greatest of these were the temples of Abu Simbel built by Ramses II (1279–13BC). The four colossal statues of Ramses in front of the temples are outstanding examples of ancient Egyptian art. The temples, carved out of a sandstone cliff on the West Bank of the Nile, were unknown in the West until their discovery by Johann Burckhardt in 1813. During the construction of the High Dam, the Egyptian government and UNESCO cooperated in a massive scheme to move the temples 200ft (60m) further up the cliff to avoid them being submerged by the rising waters.

In 2002 the center won the prestigious Emporis Skyscraper award for "World's Best New Skyscraper."

The tower's silver reflective glass cladding covers 915,000sq ft (85,000sq m).

It took three years to go through 100 architectural submissions before finally deciding on the design.

Kingdom Center

For years, Saudi Arabia's affluent middle classes have been flying to London, New York, and Paris for their shopping, but now the citizens of Riyadh have their own magnificent shopping mall, Al Mamlaka or "The Kingdom." Al Mamlaka sits at the base of a towering new development called the Kingdom Center. This was the brainchild of Saudi Prince Al-Waleed bin Talal bin Abdulaziz Al-Saud and was completed in 2001 at a cost of almost $458 million. In a city where the average height of a building is no more than five stories, the elegant Kingdom Center–a joint design project between the Omrania Corporation of Riyadh and Ellerbe Becket of Minneapolis–stands out like a huge, gleaming needle thrust point first into the desert.

The skyscraper, a silvery, reflective, oval glass tower, is intended to be seen as a global icon of Riyadh, much as the Eiffel Tower is in Paris. (In fact, they share the same height of 984ft/300m). The facade appears silvery blue in daylight and gives no indication of the number of floors within. The top third of the tower is purely decorative (city regulations state that only 30 floors can be occupied), and it is this top third that sets the skyscraper apart from others. It is an inverted parabolic curve, the ends of which are connected by a glass-clad observation deck, and it looks like the eye of some giant needle. This observation deck provides unparalleled views of the city.

WOMEN ONLY

With Saudi Arabia's strict interpretation of Islamic law it is generally not possible for women to try clothes on when they go shopping. And shops are traditionally staffed by men. Now, in the Kingdom Mall, there is a "Ladies Only" third floor. Here, veils can be left at the door, as the floor is staffed entirely by women and men are not allowed in. In fact, one of the complex's three vehicle entrances leads straight to the women's level.

Prince Al-Waleed

"If I'm going to do something, I do it spectacularly or I don't do it at all." So spoke Prince Al-Waleed (b. 1954) to the *Wall Street Journal* on the completion of his grand Kingdom Center scheme. The Prince, who is the grandson of the founder of Saudi Arabia and nephew to the present King Fahd, has cut a swathe through the modern business world. In 1991 he bought a 15 percent share of the giant US banking corporation Citicorp, quickly turning an $800 million stake into $2 billion. Prince Al-Waleed became a billionaire by the age of 31, and by 1999 *Forbes* magazine had ranked him the second richest businessman in the world after Microsoft founder Bill Gates.

East Meets West

The large, airy, atrium contains freestanding glass enclosures (inspired by the old chests carried on camels by travelers when crossing the desert), some three stories high, which house various businesses. The atrium itself is a modern interpretation of the old Middle Eastern and Central Asian tradition of the caravanserai, an inn surrounded by a courtyard where weary travelers could rest for the night. Stores line the edge of the atrium and include British department store Debenhams, and a branch of the well known American store Saks Fifth Avenue. The entire complex also includes a 225-room, five-star hotel, Prince Al-Waleed's business headquarters, and centers for weddings and conferences.

Shaped like a billowing sail to celebrate the maritime history and past dhow trade of the United Arab Emirates, the Burj al Arab Hotel is as superior to the average high-rise tower as the Chrysler Building is to other New York skyscrapers, only more so. Although built, quite literally, on sand, it rests securely on 250 reinforced concrete columns that reach 148ft (45m) beneath sea level to solid bedrock. A short causeway joins the island to the nearby mainland, reinforcing the sense of privacy and exclusivity the Burj al Arab Hotel seeks to promote and that its guests demand.

B urj is an Arabic word meaning "tower" which was originally generally applied to a fortified tower standing alone in the desert or dominating a hillside above a medieval town. But there's nothing medieval about Dubai's Burj al Arab, or Tower of the Arabs, and the only thing likely to keep people out is the cost. However, if you have the financial wherewithal, there can be few places on earth more welcoming than this famous hostelry, the only seven-star hotel in the world.

A City Icon

Paris has the Eiffel Tower, Sydney the Opera House, London Big Ben and San Francisco the Golden Gate Bridge, and so Dubai, once no more than an undeveloped and impoverished fishing port, required iconic representation, or so its government decided, once oil riches had transformed the economy. That icon is the Burj al Arab.

Burj al Arab Hotel

the facts

The tower, 1,053ft (321m) high, with 60 stories, was the 19th tallest building in the world in 2006.

In 2006 the Burj al Arab had the highest atrium in the world; at 596ft (182m) it rises to a triangular blue ceiling.

The building is illuminated at night, and the lights are switched on at 7pm. Every 15 minutes the color scheme changes automatically.

The mast on top of the building is 341ft (104m) tall.

TRADITIONAL WIND TOWERS

Air conditioning may now be the usual way to keep cool in Dubai, but in times past the severe summer heat was best combated by traditional wind towers, until just a few decades ago the tallest structures to be seen in the United Arab Emirates. These effective and environmentally friendly cooling devices consist of tall towers with four concave faces so that, wherever the wind is blowing from, one side will always capture the breeze. The base of the tower is open to permit these breezes to provide cooling relief to those in the building beneath.

Wind speed increases quite markedly with height, so the wind towers are built high. In especially hot weather, damp cloth is hung in the airflow to achieve an evaporative cooling effect. Wind towers are generally painted white, and used to be a characteristic feature of the Dubai skyline. Today they have all but disappeared from the city, but in small oases wind towers looming amid the palms are still a feature, and steps are being taken to preserve those that survive as part of Dubai's cultural heritage.

LIGHTING THE DARKNESS

Burj al Arab has the largest and most complex architectural lighting system ever installed in a single building. There are 14,000 dimming circuits distributed throughout the 202 suites as well as lighting in corridors, balconies, and public areas, including the giant atrium. Every suite has one or more dimming controls to operate the lighting in each room, with the largest suites having five systems, giving a total of 160 channels of lighting. Dimmer racks controlling all the landings are networked together by computer and operated by a custom graphic interface.

All suites are fitted with full computer facilities, as well as fax machines and 42-inch (106cm) television screens. The entire structure is air conditioned in both public and private areas. As a consequence, the building has the energy requirements of a small city, and is fitted with back-up generators to provide complete independence in the event of a grid failure. The exterior of Burj al Arab is illuminated every night by computer-controlled displays that change regularly, creating a tableau of colors on the building's facade.

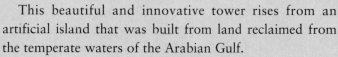

This beautiful and innovative tower rises from an artificial island that was built from land reclaimed from the temperate waters of the Arabian Gulf.

The Burj al Arab is covered with a membrane facade made of Teflon. This is brilliantly white during the many hours of sunshine, but provides a perfect background for the constantly changing computer-generated light show that is put on every night. As befits such an exclusive a hotel, there are no ordinary rooms, only 202 double-floor suites. Each is sumptuously decorated, with windows that stretch from floor to ceiling offering stunning views across nearby Dubai City and the Arabian Gulf. Every suite has its own private butler, so it is hardly surprising that occupancy rates start at around $1,000 per night, while the superior "Royal Suites" cost a regal $5,000 per night.

Visitors to the Burj al Arab are often flown in by helicopter from Dubai International Airport to the helipad, which is spectacularly situated 695ft (212m) above the ground. There is also an underwater restaurant, the Mahara Seafood, reached by a short submarine ride from the hotel lobby. Seated diners have a close-up view of the teeming marine life of the Gulf through a long, curving wall of reinforced glass.

In one way the hotel does resemble a mediaeval Arab fortress, as access to the hotel is only possible by reservation or invitation. Previously, non residents could buy a pass simply to view the lobby of the hotel, but this is no longer possible. A strict dress code is also enforced, with casual clothes such as sneakers and shorts unacceptable.

More than 11.5 billion cu ft (325 million cu m) of sand was dredged to build the islands.

Property values at The World more than doubled between 2003 and 2005.

Together with The Palms, The World has added more than 400 miles (640km) to the Dubai shoreline.

The World

Like The Palms in Dubai, The World is another huge building project that has taken shape in the shallow waters of the Arabian Gulf, some 3 miles (5km) off the shore of Dubai. And like The Palms, it was conceived by Sheikh Mohammed bin Rashid Al Maktoum, Crown Prince of Dubai.

The World comprises a collection of manmade islands grouped within an oval breakwater to form the shapes of the continents of the world. It covers an area 6 miles (10km) by 4 miles (6km), including 1,109sq yards (927,300sq m) of beach. Each of the 250 to 300 different sized islands are themed to reflect different countries, and the properties on them are made up of exclusive private homes, estates, and luxury resorts. Golf courses, watersports, hotels, marinas, and even an African game reserve are just some of the leisure facilities and attractions on offer. Access to the islands is by boat only.

More to Come

The World is part of Dubai's strategy to become a Middle Eastern business hub. More controversially, the tiny but liberal emirate seems to be considering giving Las Vegas a run for its money as an entertainment center. The project is also in line with the Dubai government's continued efforts to boost tourism, which is already growing at an incredible rate. Visitor numbers are predicted to reach 12 million by 2007 and 42 million by 2015.

After The World, Dubai has plans to construct a still larger project, the Dubai Waterfront. This will redevelop the emirate's last stretch of untouched shoreline and include a city housing 400,000 people on a series of manmade islands and canals.

CROWN PRINCE OF DUBAI

In the last 12 years Dubai's nominal gross domestic profit has risen dramatically from $8 billion to $20 billion, which has largely been due to the Crown Prince of Dubai's tireless enterprise. Sheikh Mohammed (b. 1949), also the United Arab Emirates' defense minister, is the *de facto* ruler of Dubai as his elder brother spends most of his time in the United Kingdom. Apart from his support of major projects such as The World, the Burj al-Arab Hotel and the soon-to-be tallest building in the world, the Burj Dubai, the Sheikh also has a deep love of horses and hosts the world's richest horse race, the Dubai World Cup.

LIMITED OIL RESERVES

Unlike neighboring countries such as Saudi Arabia, Kuwait, and Iraq, Dubai has limited oil reserves. Oil was discovered offshore in 1966 and production commenced in late 1969. In 1991 Dubai's oil reserves were estimated at 4 billion barrels, but it is predicted these will be used up by 2016 if current levels of production continue. The government is well aware of the implications of this and is searching for alternative sources of prosperity, epitomized by projects such as The World.

- More than 1,590,000cu ft (45,000cu m) of concrete has been poured into the foundations.
- The 700 private apartments in the building were sold within eight hours of going on the market.
- Dubai Mall covers an area slightly more than that of 50 soccer pitches.

Burj Dubai

These days Dubai appears to be awash with ambitious schemes and the Burj Dubai (Arabic for Tower of Dubai) might just be the biggest of them all. The super-skyscraper, anticipated to be finished in 2008, will be the world's tallest ever structure. Its finished height is estimated to be 2,314ft (705m), but there have been suggestions that it may go as high as 2,640ft (800m). Formerly, the tallest manmade structure was the Warsaw radio mast at 2,120ft (646m), but this collapsed in 1991 while being renovated.

The developers have refused to disclose the exact height of the Burj and are prepared to make the tower even higher during construction if it appears any of their rivals around the world intend to top them. Contenders include New York's proposed Freedom Tower, set to replace the Twin Towers destroyed in 2001.

Groundbreaking work began on the project in January 2005 and has proceeded apace. There are expected to be 160 floors above ground, but the number below ground is unknown at present. The US firm of architects Skidmore, Owings & Merrill drew inspiration for the tower's design from the six-petal desert flower of the region. Rising out of an artificial lake, the silver steel and glass tower appears to spiral its way to a dramatic point.

A Home in the Clouds

The Burj Dubai is first and foremost a residential building, with floors 45 through to 108 being private apartments. A hotel will occupy the first 37 floors, with the remainder of the building given over to offices. There will be an observation deck on the 124th floor with an open-air terrace for anyone brave enough to face the strong winds.

The Burj will sit at the heart of a large, mixed-use development. On site will be the world's largest shopping center, the Dubai Mall, which will include, amongst other attractions, a gold *souk* (market) and an ice-rink. Also part of the complex will be a brand new "Old Town," laid out in the style of the Al-Bastakia quarter in Bur Dubai.

BURJ DUBAI

A PASSION FOR TALL BUILDINGS

Burj Dubai architect Adrian Smith of Skidmore, Owings & Merrill has had a hand in many of the world's tallest structures, including the John Hancock Center in the United States and China's tallest building, the Jin Mao Tower in Shanghai. Smith is a firm believer in buildings fitting their time and place, and, crucially, always being culturally relevant. In North American cities, with their history of skyscrapers, this has never posed a problem, but in Dubai, where few buildings have ever risen above five stories, the significance of such iconic structures has yet to be established.

THE GREAT PYRAMID AT GIZA

For a time, the Burj Dubai will be the tallest structure on earth, returning the Middle East to a position it has not held since Lincoln Cathedral in England topped the Great Pyramid at Giza early in the 14th century. For 38 centuries the Great Pyramid, rising to an original height of 481ft (146m), stood proudly above anything man had ever previously achieved. It was built to house the mortal remains of the fourth dynasty Egyptian Pharaoh Khufu, perhaps better known by his Greek name, Cheops. The Great Pyramid took an estimated 20 years to build, while Burj Dubai will take just three years.

159

Palm Deira measures 9 miles (14km) from the base of the trunk to the top and 5 miles (8km) across.

Construction of Palm Deira began at the end of 2004 and is expected to be completed within five years.

Palms Jumeirah and Jebel Ali are comprised of 100 million cubic yards (130 million cu m) of rock and sand.

The Palms

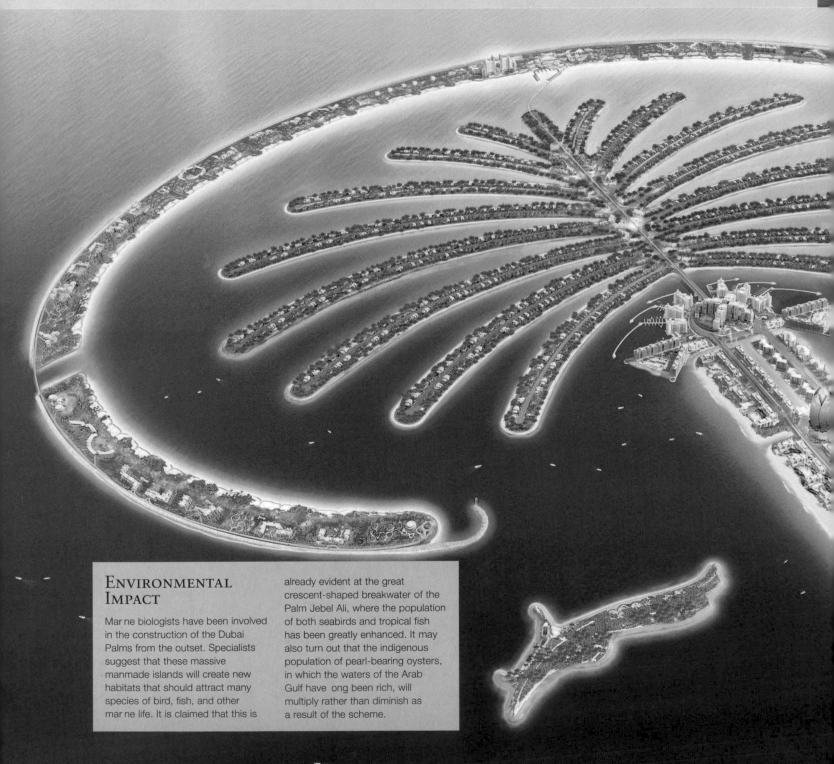

ENVIRONMENTAL IMPACT

Mar ne biologists have been involved in the construction of the Dubai Palms from the outset. Specialists suggest that these massive manmade islands will create new habitats that should attract many species of bird, fish, and other mar ne life. It is claimed that this is already evident at the great crescent-shaped breakwater of the Palm Jebel Ali, where the population of both seabirds and tropical fish has been greatly enhanced. It may also turn out that the indigenous population of pearl-bearing oysters, in which the waters of the Arab Gulf have ong been rich, will multiply rather than diminish as a result of the scheme.

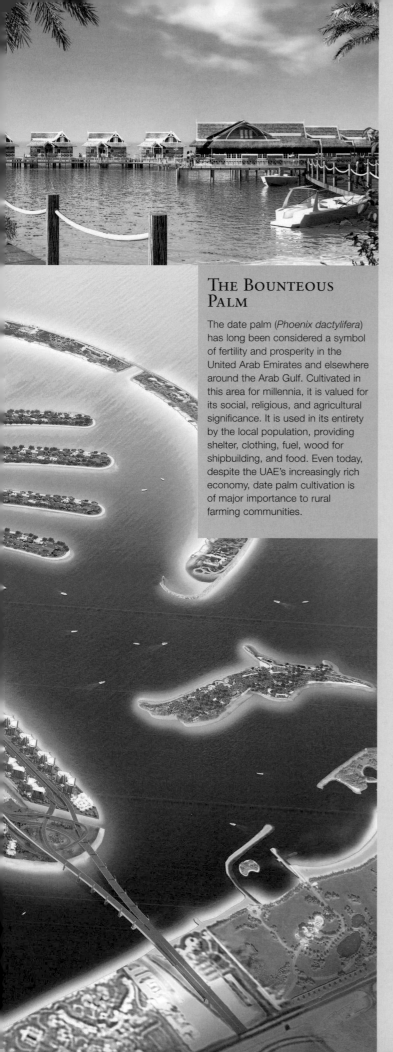

THE BOUNTEOUS PALM

The date palm (*Phoenix dactylifera*) has long been considered a symbol of fertility and prosperity in the United Arab Emirates and elsewhere around the Arab Gulf. Cultivated in this area for millennia, it is valued for its social, religious, and agricultural significance. It is used in its entirety by the local population, providing shelter, clothing, fuel, wood for shipbuilding, and food. Even today, despite the UAE's increasingly rich economy, date palm cultivation is of major importance to rural farming communities.

Best appreciated from the air because of their huge size, the Palms at Dubai are astonishingly lavish (and profitable) property developments quite unlike anything else. They were conceived and built on the instruction of Sheikh Mohammed bin Rashid Al Maktoum, Crown Prince of Dubai. He was aware that oil reserves in the United Arab Emirates (UAE) are limited, while supplies of sand and rock are virtually limitless. Another factor which lead to the project going ahead was that the warm, shallow waters of the Arabian Gulf off the shores of Dubai are protected from the worst ocean storms by the narrow Strait of Hormuz that links the Gulf with the wilder waters of the Arabian Sea.

Two manmade "islands," Palm Jumeirah and Palm Jebel Ali, each take the shape of a gigantic date palm, with the base of the trunks connected to the mainland by 990ft (300m) bridges. The trunks are crowned with 17 cascading fronds, and crescent-shaped islands act as breakwaters. The project will add 72 miles (120km) of coast to the Dubai shoreline, as well as creating huge new residential, leisure, and entertainment areas.

Palm Jumeirah is divided into three areas. The outer crescent is for luxury resorts and themed hotels, for example replicating traditional elements of Japanese, Brazilian, and Italian culture and architecture. The fronds are secluded private neighborhoods with luxurious Arab- and Mediterranean-style villas, complete with swimming pools and private beaches. The trunk is the access hub with theme parks, marinas, shopping malls, and restaurants. Palm Jebel Ali is similar, but is 50 percent larger.

The Third Tree

The project has been so successful that a decision has been made to construct a third palm, Palm Deira. This development, the largest of the three, will have an overall area of 31sq miles (80sq km)—larger than Manhattan. Palm Deira will follow the same concept with a trunk, a crown of fronds (41 on this tree), and an outer crescent. The crescent will run for 13 miles (21km) and will be the largest manmade breakwater in the world. To complete the project 1.3 billion cubic yards (1 billion cu m) of rock and sand will be needed.

- Before the university was founded, Omani citizens had to go abroad if they wanted to go to study.

- Between 1986 and 2005 the number of students has grown from 500 to more than 10,000.

- Sultan Qaboos University Hospital is a 500-bed facility for teaching medicine.

the facts

Sultan Qaboos University

Sultan Qaboos University, located at Al Khoudh, some 30 miles (50km) west of Muscat on the undeveloped southeast coast of the Arabian Peninsula, was conceived and founded by Sultan Sayyed Qaboos. Work began in 1982 and in 1986 the first students were enrolled at the five colleges: Medicine, Engineering, Agriculture, Education and Science. In 1987 the College of Arts was opened, followed by the College of Commerce and Economics in 1993.

Islamic Architecture

The university lies in a valley of stark beauty beneath the arid Oman Mountains. Constructed in white and pink sandstone, with arches and courtyards, its style is designed to reflect both local Omani architecture and the wider Islamic tradition. The campus is laid out along an axis aligned with Mecca in Saudi Arabia, that starts at the massive, traditional wooden gates of the university. From there the axis runs through the colleges and administrative buildings to the University Mosque at the western end of the campus. The mosque stands on higher ground and is visible from all over the university.

SULTAN QABOOS

Regarded by most of his subjects as a benevolent ruler, Sultan Sayyed Qaboos ibn Sa'id Al 'Bu Sa'id (b. 1940) has been responsible for the Sultanate of Oman's entry into the modern world. Since coming to power in 1970, he has used the country's oil revenues to fund vital infrastructure projects. Before this, Oman was one of the poorest countries in the Middle East. Education became a rallying cry for the Sultan and by the first five months of his rule, there had been a 662 percent increase in the number of children receiving an education. Girls are now also allowed to attend school.

AL AZHAR UNIVERSITY

The Arab world has a long tradition of higher education, dating back almost to the beginning of the Islamic Era in AD622. Al-Azhar University in Cairo was founded in 988 and is considered one of the oldest functioning universities in the world. The library holds more than 99,000 books comprising 596,000 volumes of precious manuscripts, some dating back as far as the 8th century.

Verdant Gardens

Considerable care was applied to the landscaping of the grounds, which were planted with species of trees and plants native to Oman and the Gulf. The plan draws its inspiration from traditional Omani hill towns, with defensive walls and secret gardens unfolding within them. These verdant gardens provide welcome shade, and shelter against the wind and dust storms that sometimes affect the region. Although a revolutionary concept for Oman, the university remains conservative by international standards. Men and women have separate access routes to colleges, and there is separate seating in the classrooms. However, female students form about half of the total population.

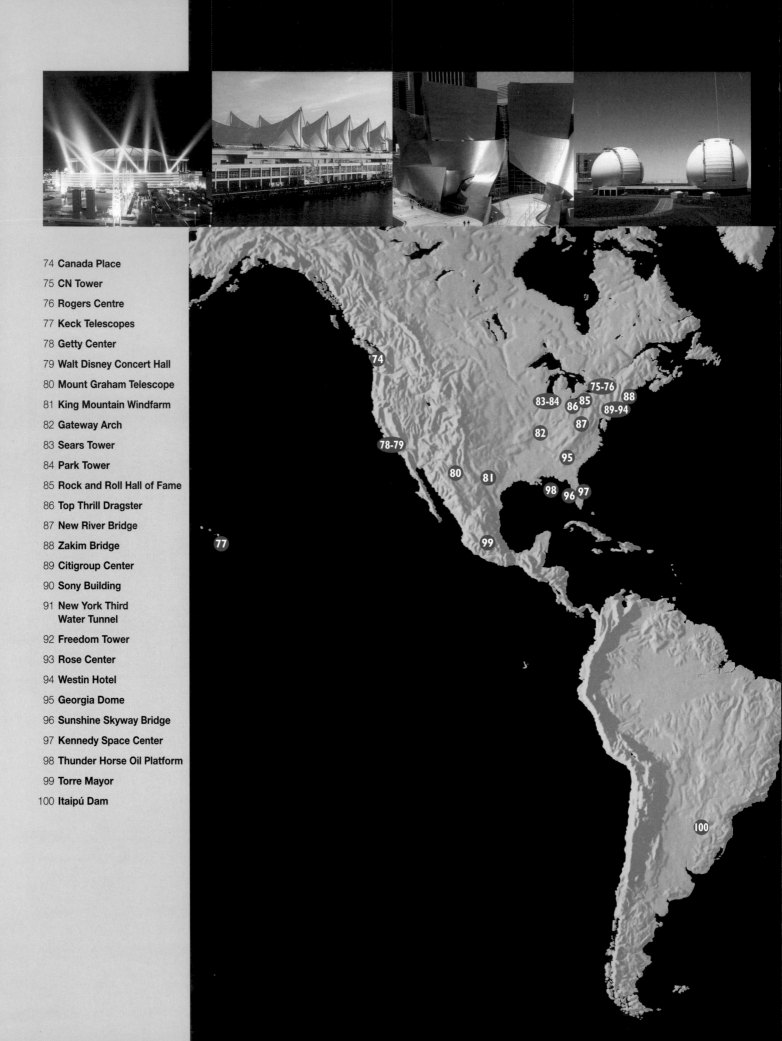

The Americas

T he skyscraper is the USA's most recognizable contribution to world architecture, and Chicago is credited as the birthplace of the modern skyscraper, (although buildings many stories high were being built in Yemen in the mid-East several centuries ago). Classic Chicago skyscrapers such as the Park Tower and the Sears Tower are included in this chapter. But north and south America have contributed far more than just tall buildings to the landscape—the USA is at the forefront of design in areas as diverse as telescopes, windfarms, bridges, and even rollercoasters.

The Kennedy Space Center in Florida is a prime example of American innovation. Here the sheer size and scale of the Vehicle Assembly Building, which is one of the largest buildings in the world, is breathtaking. For example, its doors are three times as high as the Statue of Liberty and take an hour to open. Similarly large-scale projects include the Thunder Horse oil platform in the Gulf of Mexico and New York's Third Water Tunnel, a major engineering project which is still ongoing.

In terms of buildings which are memorable for their striking beauty, Gehry's Walt Disney Concert Hall in Los Angeles is just that. The swirling asymmetric curves, which are cloaked in stainless steel, reflect the almost constant sunshine of Los Angeles, and provide a focus for the city's notoriously formless center. Looking to the future, the most eagerly anticipated new building is New York's Freedom Tower, which will stand on the site of the World Trade Center's twin towers. Its gleaming, crystalline skyscraper, in the form of an obelisk, is due to be completed by 2010.

The five tented roofs, made from Teflon-covered fiberglass, are suspended from steel masts.

Construction began in March 1983 and was completed by December 1985.

In 2005 the center began an expansion program due for completion in time for the 2010 Winter Olympics.

Canada Place

EXPO '86

This world trade fair was held in Vancouver to celebrate the 100th anniversary of the city's foundation. The theme was transportation and communication, with exhibits in the Canada Pavilion including a space-age airship and a prototype Maglev train powered by magnetic levitation. In all, 54 nations participated in the fair and visitor numbers reached 22 million. Despite a large deficit of more than US$300 million the event proved a huge boost to Vancouver's image and economy, besides leading to regeneration of the waterfront. Most of the 175-acre (71ha) site has since been developed for housing or turned into parkland.

Only in recent times have the waterfronts of the world's great coastal cities shed their practical, workaday image for one of chic living. Vancouver is no exception. Canada's third-largest city occupies a spectacular location on one of the world's great natural harbors, yet until the 1980s the downtown area was cut off from the sea by a dreary zone of long-abandoned timberyards, rusting rail-tracks and rotting wharfs. Today the scene is very different, with stupendous sea-views now revealed and the old waterfront reinvented to reflect the needs and dreams of modern urban life. And, just as is the case in Sydney or Bilbao, an iconic building stands as a symbol of this transformation and provides a focus for the whole breathtaking scene. Thrusting out into the harbor, Canada Place's unique, futuristic form evokes the image of a great ocean schooner under sail.

A 23-story tower represents the bridge, promenade decks run around the long, low-slung hull, while the five tall roofs of tented fabric that soar above the central superstructure are quite clearly intended to be seen as sails.

Rejuvenation and Redevelopment

Since its completion in 1986 the complex has, like the waterfront in general, undergone changes and developments. It was originally designed as the Canada Pavilion for Expo '86, but, unlike the temporary structures on the main site of the fair just across the water, it was always intended to become a permanent feature of the city that would serve a whole range of useful roles. In keeping with its ship-like form, the lower levels provide a spacious terminal for cruise-liners sailing to Alaska. The high-rise tower houses a 504-room luxury hotel, while the main pavilion is now a conference and exhibition center that can host major international events. With an IMAX movie theater, a ballroom and an amphitheater, the complex also plays a lively role in terms of public entertainment, as well as offering terraced promenades where people can stroll.

GOOD FOUNDATIONS

The site of Canada Place was formerly known as Pier BC, a huge, disused wharf that belonged to Canadian Pacific Railways. Its 6,000 concrete piles, dating from 1927, were so well made that they required only minor restoration. About 1,050 new piles were added, along with concrete-filled steel caissons that were sunk as foundations to depths of up to 180ft (55m). A further 1,000 piles will be needed for the new extension, which will enlarge the complex by 134,000sq yards (112,000sq m).

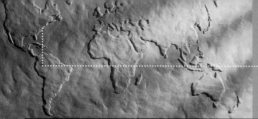

- The tower contains 53,000 cubic yards (40,524cu m) of concrete and weighs 130,000 tons (118,000 tonnes).
- The record for climbing the 1,776 steps of the world's longest staircase is 7 minutes 52 seconds.
- CN Tower stands for the tower's full name of Canada's National Tower.

CN Tower

TALLEST BUILDING

Over the years, the CN Tower has been given the titles of World's Tallest Building, Tallest Tower, and Tallest Freestanding Structure, and since 1976, no freestanding structure or building has surpassed its height. However, definitions of what constitutes a "building" have become increasingly confused since the Empire State Building broke all records in 1931. The height of a building's topmost floor, its roof, any spire or other architectural detail, or even the tip of its antenna, are now all likely to be cited when rival claims are considered. The argument looks likely to be resolved once and for all by the completion of the Burj Dubai, which has been designed to be unquestionably the tallest structure in the world. It is expected to reign supreme on all accounts for a long time to come.

In 2006 the CN Tower, at 1,815ft (553m) high, remained the tallest self-supporting structure in the world, a title it has held since it was completed in 1975. From the start it was important that Toronto's famous landmark should be of record-breaking size. In the late 1960s, newly built skyscrapers, some up to 800ft (245m) high, were wreaking havoc with radio and television reception in the city. Broadcasting companies all agreed that a shared, super-tall communications mast was needed, but what might have been a simple engineering project took on a new twist when a visionary local architect, Ned Baldwin, persuaded Canadian National Railways to invest in a far more ambitious scheme. The CN Tower that Baldwin proposed would not only provide Toronto with crystal-clear radio and TV signals, but it would also be a source of national pride. As the world's highest "building," it would attract admiring visitors to its observation decks and restaurants, while its space-age design would demonstrate the skills of Canadian engineers.

Don't Look Down

Once construction started in February 1973, work proceeded at a hectic pace, with crews working 24 hours a day, five days a week. Even now it is possible to see the lines left each week when the concrete was left to set over the weekend. In August 1974 work started on the seven-story Main Pod that houses the broadcasting equipment and the main observation galleries. Finally, in March 1975, a helicopter winched the 335ft (102m) antenna into position.

Since the CN Tower opened to the public it has attracted 2 million visitors a year, and has repaid its construction costs of CAN$63 million many times over. Now more than 30 years old, it has been given a facelift, ensuring its position as one of Canada's most popular attractions.

Six elevators take visitors up to 1,150ft in just 58 seconds, where the world's highest restaurant rotates at one revolution every 72 minutes. Still higher, at 1,465ft (447m), is the world's highest observation deck, with views of up to 75 miles (120km). For many, the most memorable view is vertically down through the reinforced Glass Floor—an experience akin to walking on air.

LIGHTNING

The tower is struck by lightning at least 75 times a year, although the charge is harmlessly diverted through metallic conductors earthed to 42 buried rods. In 1991 the SkyPod was used as a laboratory for testing ways of protecting sensitive electronic equipment, such as that used on aircraft, from the effects of lightning strikes. It was found that the tower's unusual shape influenced the charge, a discovery that may affect aircraft design.

More than 8 miles (13km) of zippers connect the strips of artificial turf.

When the roof is opening or closing, the sliding panels travel at a rate of 71ft (22m) per minute.

The video display is one of the largest in the world, measuring 33ft (10m) by 110ft (34m).

Rogers Centre

ROOF FACTS

The roof is constructed from insulated acoustic sheet-steel and each section is clad in a single polyvinyl membrane. It covers an area of 3,771sq yards (31,526sq m) and is held together with 250,000 bolts. Opening and closing the roof panels takes 20 minutes and, when closed, the apex of the dome is high enough to accommodate a 30-story building.

The Rogers Centre (formerly known as the SkyDome), on the shores of Lake Ontario in Toronto, broke new ground as the world's first multipurpose sports and entertainment building to have a fully retractable roof. Completed in 1989, the landmark structure, resembling a huge, blinking eye, was commissioned and designed as a key element in the urban renewal of downtown Toronto. It is home to the city's Blue Jays baseball team, along with the Canadian Football League's Toronto Argonauts.

Sliding Roof

The most startling feature of the near-circular building is its sliding roof, measuring some 674ft (205m) at its widest point and with a height of 282ft (86m). When required, the two central hooped roof sections mounted on tracks slide to the end of the playing field. The third section, the cap end of the massive roof, rotates and tucks under the two central sections, opening the field to the sky and giving spectators unobstructed views out of three sides of the stadium.

Inside, other innovative features include seating on tracks that enables the space to be reconfigured from a baseball field to a football stadium or concert venue in a matter of hours. The five tiers of seating, which hold a capacity crowd of 50,516 for baseball and 53,506 for football, include the 161 prestigious Sky Boxes for corporate entertaining at levels three and four. Also at these levels are a 650-seat restaurant and a 300ft (91m)-long bar. The building includes a 348-room hotel with 70 rooms overlooking the field.

Architects Rod Robbie from Toronto and Michael Allen from Ottawa designed the stadium and construction began in April 1986. The Blue Jays played the first game on 5 June, 1989 before an ecstatic crowd. In February 2005 the complex was bought by Rogers Communications, owner of the Blue Jays.

THE BLUE JAYS

A competition in 1976 to find a name for the newly formed Toronto baseball team attracted more than 30,000 entries. The winning entry suggested the blue jay, which is a brightly colored bird with a crest which is commonly seen in southern Canada. A stylized version of a blue jay, with its head extending from the letter 'J' of 'Jays,' forms the team's logo. The Blue Jays played their first game in April 1977 against the Chicago White Sox, and won the game 9-5.

Keck has a complex system that combines light from both telescopes, giving a tenfold increase in power.

If the mirrors were enlarged to the diameter of the earth, the largest imperfection would be 3ft (1m) high.

The telescopes could distinguish between a car's twin headlamps at a range of 500 miles (800km).

the facts

Keck Telescopes

ORIGINS

Since 1998 the Keck telescopes have played a central role in NASA's "Origins" program, searching space for stars with planetary systems that might support some form of life. Well over a hundred "extra-solar" planets have so far been identified, and observations are now concentrating on around 150 comparatively nearby stars (within 50 light years of our sun) in the hope of finding smaller planets with similarities to earth. A breakthrough came in June 2005 when astronomers were able to detect a planet only twice the size of earth orbiting the star Gliese 876, although its density and surface temperature ruled out the possibility of life.

For many years after the construction of the Hale Telescope on Mount Palomar, California, in 1947, astronomers believed that its 16ft (5m) mirror was the largest that could possibly be built. In 1977, however, scientists at the University of California devised an entirely new kind of reflective surface that, however large, would not deform or break under its own weight. Their discovery revolutionized telescope design and led directly to the construction of the hugely powerful Keck telescopes at the Mauna Kea Observatory on Hawaii, which were begun in 1985.

Keck I and Keck II form a matching pair of giant instruments placed 93 yards (85m) apart. Each has at its heart a 33ft (10m) mirror that, instead of being ground from a single solid disc, is made up of 36 hexagonal segments that interlock with phenomenal precision to form a single parabolic surface. Each segment, weighing half a ton, is 7ft (2m) wide by 3in (8cm) thick and is powered by electric motors that constantly adjust its angle to compensate for distortions as the telescope tilts. These adjustments, which are vital to maintain the mirror's perfect curve, are on a truly microscopic level, measuring as little as 4 nanometers, or one millionth of an inch.

Probing Time and Space

Keck I saw "first light" a term used by astronomers for a telescope's first observation, on 4 December, 1990, when, with just nine mirror segments in position, it was used to photograph NGC 1232, a galaxy 65 million light years from earth. Once Keck II was completed, the observatory became fully operational in October 1996. Weighing 298 tons (270 tonnes) apiece, the telescopes are housed in giant domes 98ft (30m) high and 118ft (36m) wide on the 13,800ft (4,800m) summit of Hawaii's highest mountain. To avoid thermal disturbance of the instruments, the temperature in the domes must be maintained at just above freezing. Although Keck is the largest of the telescopes on Mauna Kea, 11 countries have 13 different instruments on the summit, which enjoys some of the best viewing conditions in the world.

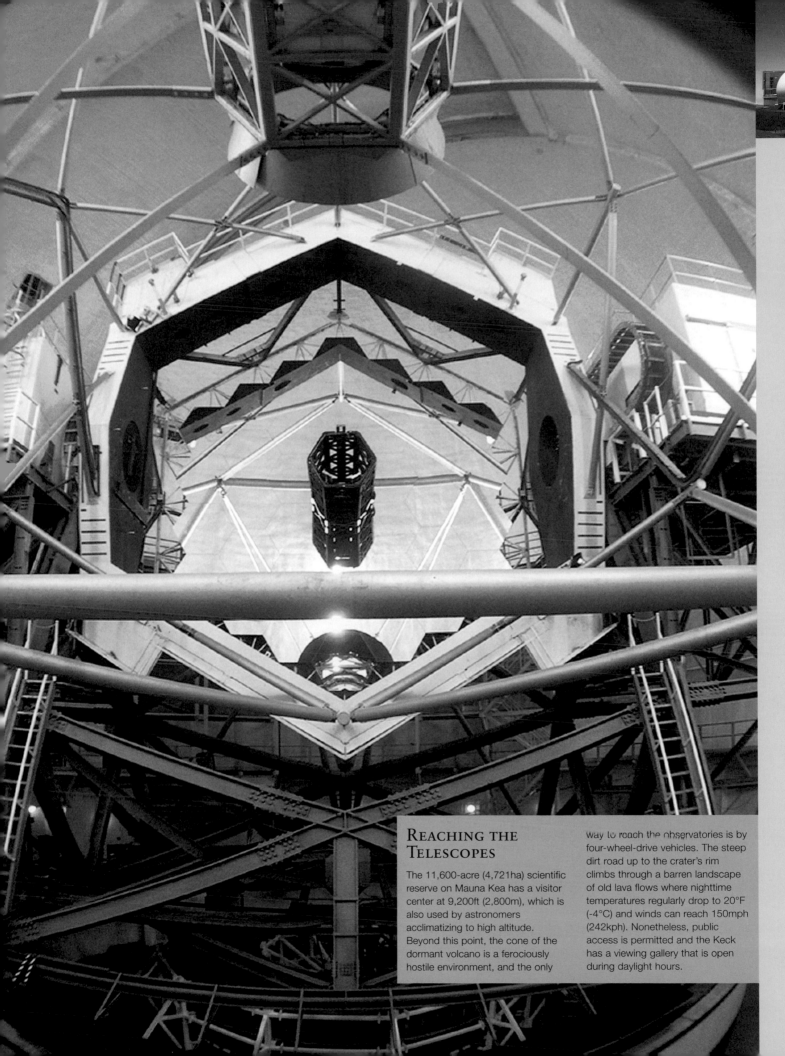

REACHING THE TELESCOPES

The 11,600-acre (4,721ha) scientific reserve on Mauna Kea has a visitor center at 9,200ft (2,800m), which is also used by astronomers acclimatizing to high altitude. Beyond this point, the cone of the dormant volcano is a ferociously hostile environment, and the only way to reach the observatories is by four-wheel-drive vehicles. The steep dirt road up to the crater's rim climbs through a barren landscape of old lava flows where nighttime temperatures regularly drop to 20°F (-4°C) and winds can reach 150mph (242kph). Nonetheless, public access is permitted and the Keck has a viewing gallery that is open during daylight hours.

- On completion of the project, the sloping sides of the hill were planted with thousands of oak trees.

- More than 500 varieties of plants grow in the Garden.

- In deference to local wishes the height of the buildings was restricted to two stories, so underground levels provide storage and connecting passageways

Getty Center

RICHARD MEIER

Richard Meier, born in Newark in 1934, is a stalwart of American modernist architecture and is known worldwide for his pure white buildings. He began his career by designing one-off houses, but soon moved on to larger corporate and public projects including large numbers of museums—his best known works include the Frankfurt Museum of Decorative Arts in Germany, the Barcelona Museum of Contemporary Art in Spain, the Siemens HQ in Munich and the Canon HQ in Japan. In 1984 Meier was awarded the Pritzker Prize for Architecture, considered the field's highest honor.

n its majestic position in the Santa Monica Mountains overlooking Los Angeles, the Getty Center has been described as a modern-day Acropolis. With its range of low-rise, stone-clad buildings arranged around squares and courtyards, this stunning project occupies an entire hilltop and rates as one of the world's largest arts complexes. The architect, modernist master Richard Meier, has described the project as growing from the topography of the area and responding to the grid of the city and the flow of the freeway. He has previously stated that the design and construction of the center, which opened in 1997, totally consumed his life.

Meier chose to build the complex from Travertine marble, which was brought from Bagni di Tagoli, just east of Rome, in Italy. Many of the exterior walls are glass, to allow as much natural light in as possible, but lightfall is carefully controlled with louvers and filters to ensure no damage is caused to the artworks inside.

Visitors arriving by car must park at the base of the hill and take the silent, computerized tram up the steep hillside. At the top is the large Arrival Plaza with views through the complex and beyond, an auditorium and staff buildings to the left, and a restaurant and a café to the right. A long flight of steps leads up from here to the entrance to the museum—a series of five galleries in linked pavilions around a central courtyard. Organized in chronological order, the main painting galleries are at first-floor level. The ground-floor galleries show off sculpture, manuscripts, drawings, and some antiquities.

Other buildings on the site house the Getty Conservation Institution, the Getty Center for Education, the Getty Grant Program, and the Getty Research Institute for the History of Art and the Humanities. East of the museum in a natural ravine is the Central Garden, created by artist Robert Irwin.

HIGHLIGHTS OF THE COLLECTION

The Getty collection focuses on pre-20th-century works, including European paintings, drawings, sculpture, illuminated manuscripts, and decorative arts. There are also European and American photographs. Italian Renaissance highlights include panels by Carpaccio and Masaccio along with the *Adoration of the Magi* by Mantegna. Flemish and Dutch masterpieces from the 17th century include works by Rubens, Jan Breughel, and Rembrandt's *St Bartholomew*. The permanent collection of French decorative arts is particularly impressive and includes fine furniture along with a handful of exquisite 18th century paneled rooms.

There are works by Impressionists and Post-Impressionists including Monet, Pissaro, and Van Gogh, whose *Irises* is perhaps the best-known piece in the collection. The main body of Getty's Greek and Roman collection is on show in the former Getty Museum, the Getty Villa at nearby Malibu.

Walt Disney Concert Hall

Clad in a dazzling skin of stainless steel, the swirling, asymmetric curves of the Walt Disney Concert Hall perform a wild dance above the rooftops of downtown Los Angeles. Designed by Frank Gehry, an architect famed for his unconventional ideas, it is a building of fantastical exuberance that has been as carefully composed as any orchestral score. Indeed, the conductor of the L.A. Philharmonic, Esa-Pekka Salonen, has described the hall's extraordinary architecture as "the most beautiful frozen music of our time."

I t was a $50 million gift by Lillian Disney, the widow of Walt Disney, which first put the project on the road in 1987. But, throughout the 15 years it took to construct, the building attracted controversy and debate. There were problems with changes in design, rows with city government, and spiraling costs. It was not until Gehry's equally outrageous Guggenheim Museum in Bilbao received a rapturous international reception on its opening in 1997 that fears were allayed. Further donations from the Disney family and immense local enthusiasm then ensured that the hall would be completed. At long last, a spectacular Hollywood-style opening was held in October 2003.

Since then Gehry's masterful building has attracted admiration, not only as a music venue, but also as a landmark that provides the notoriously formless center of Los Angeles with a vibrant, visually astonishing new heart. Aside from the concert hall, the complex also includes an outdoor

More than 30,000 architectural drawings were made between the initial sketches and the final plans.

The building incorporates approximately 12,000 pieces of steel, each of individual size and shape.

The hall's final cost was $274 million, $85 million of which was provided by the Disney family.

In March 2005 parts of the stainless-steel cladding were sanded down, following complaints from local residents about reflected heat and glare.

CLAY AND COMPUTERS

Gehry's complex architecture would be impossible to realize without help from advanced computer software. CATIA (Computer-Aided Three-dimensional Interactive Application) was originally developed by the French aerospace industry for the design of jet fighters and can be used to create "virtual structures" based on mathematical co-ordinates similar to the grid of a map.

The first stage in the process consists of hand-drawn sketches by the architect. Using these sketches as a guide, skilled assistants then construct a model of the structure out of clay or plastic. With the help of a laser-pen, the precise co-ordinates of every plane and angle are then uploaded into the computer program to create a 3-D image that can be studied, altered, and perfected down to the finest detail. Not only can the engineers design the steel frame that underpins the building's curves and estimate the materials required, but the sophisticated program will even work out the likely labor costs and most efficient building schedule.

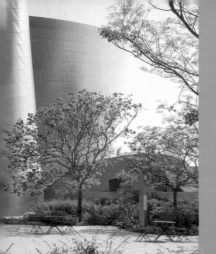

PITCH PERFECT

Aside from its fascinating architecture, the concert hall is also noted for its excellent acoustics. This is due to the fact that Gehry worked in close collaboration with a leading Japanese acoustician, Yasuhisa Toyota. To begin with, the artist and the technician were in opposition, with Gehry dedicated to sculpted forms that took no account of acoustics, while Toyota favored a strict, symmetrical design. Nonetheless, this creative opposition eventually led toward a brilliantly successful compromise.

More than 40 small models of differing shape and form were constructed, and each was carefully considered in relation to sound quality before a final choice was eventually made. This was then reproduced as a detailed 1:10 scale model that was tested with sound impulses at 10 times normal frequency to recreate the conditions of a miniature concert.

Against all expectations, the elaborately convoluted form of Gehry's hall actually intensified the richness and texture of the sound. Even the rippling timber ceiling, with its innumerable nooks and crannies, adds a warm resonance that is sometimes lacking from the music to be heard in concert halls which have a more conventional design.

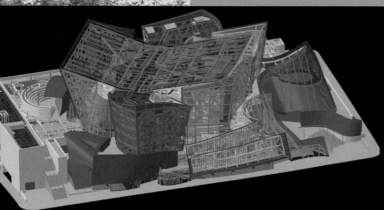

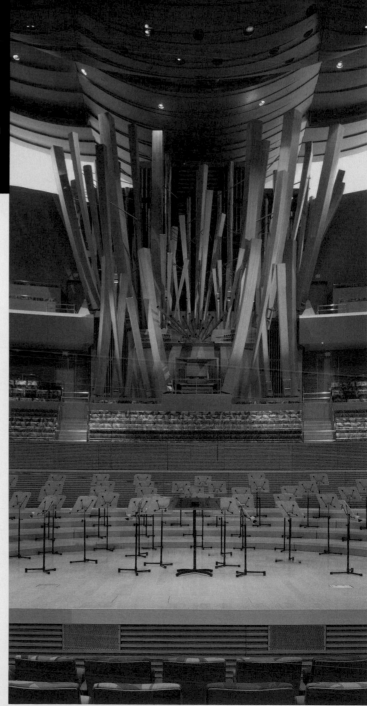

amphitheater, and extensive gardens that are shared as a "living room" by the city's diverse population.

A Symphony of Steel and Wood

In contrast to the panels of titanium that he used in Bilbao, Gehry chose stainless steel to cloak the hall's exterior in order to reflect the brilliant, almost constant sunshine of Los Angeles. The precision-jointed sheets look seamless, having been designed and cut using computers and lasers, but the structure was erected by steel-riggers who used skills that have scarcely altered since the first skyscrapers were built more than a hundred years ago.

Inside, the 2,265-seat concert hall is an intoxicating mix of natural materials and unconventional design. Lined with golden-hued Douglas fir and featuring a billowing, carved ceiling, it is a far cry from the austere shoebox form found in many modern halls. Gehry even helped design the organ that rears up behind the stage in a sculptural explosion of angled wooden pipes. Also (and unusually for a concert hall), natural light floods in through a 36ft (11m)-high window and skylights. In its use of sensuous, hand-crafted forms and luxurious materials, the hall is intended to evoke a magical barge on which audience and musicians can all voyage through the world of music. Regardless of these intentions, few can doubt that the hall is a stunning piece of modern architecture.

The 2,000-ton (1,814 tonne) rotating cube can turn at 2 degrees per second.

The structure can withstand 140mph (225kph) winds and has snow-melting heaters built into its roof.

The first mirror was delivered in 2003 after a three-day, 29-mile (46km) ascent up the unmetaled road.

Mount Graham Telescope

Close to the summit of Mount Graham, a 10,720ft (3,268m) mountain in Arizona, a strange, robotic structure rises from a forest of spruce trees. The huge steel prism standing on a concrete plinth houses one of the world's most powerful and sophisticated telescopes, which started operating in 2006. Unlike the Keck telescopes in Hawaii, the University of Arizona's Large Binocular Telescope (LBT) is equipped with 26ft (8m)-diameter mirrors that are ground from discs of solid glass rather than composed of segments. Also, again unlike the Keck, the telescope has been constructed as a single instrument, with twin lenses placed 13ft (4m) apart but sharing a common mount.

Thanks to its unique design, the LBT combines enormous power with an unprecedented quality of definition, which produces images 10 times sharper than those of the Hubble Space Telescope. Since plans for the telescope were first conceived in the 1980s, the project has developed as a truly international venture, with a consortium of scientific institutions from the USA, Italy, and Germany providing both finance and technical expertise. As part of a new generation of giant telescopes that is currently coming into operation around the world, the telescope on Mount Graham will enable astronomers to probe back through 14 billion light years to the dawn of the universe and to study planetary systems around distant stars, revolutionizing our knowledge of deep Space.

The Steel Cube

The 82ft (25m)-long telescopes are housed within a steel cube that rotates around a concrete drum above the level of the trees. Two sets of L-shaped shutters, 33ft (10m) wide, disappear into the structure's front and roof whenever the telescope is used, and four sets of ventilation doors ensure an even temperature within. Aside from the telescope itself, the cube contains engine rooms, an observation level, a visitor gallery, elevators and a crane, while laboratories and offices are housed in the plinth and in an adjoining annex.

CONSERVATION AND CONTROVERSY

From the start, the project was beset by controversies and delays. Environmental groups bitterly opposed the felling of trees in an area designated as a National Forest and claimed that the habitat of a rare subspecies of red squirrel would be irreparably harmed. Opposition also came from the San Carlos Apache tribe, which reveres the mountain as a sacred site—Dzil nchaa si an—dedicated to ancestral spirits.

Even when all of the legal challenges had been resolved, the building program faced an unexpected threat from forest fires, the last of which, in July 2004, burnt 30,000 acres (12,000ha) and came within half a mile (1km) of the observatory.

MAKING THE MIRRORS

The telescope's two primary mirrors, each of which weighs 18 tons (16 tonnes) and is worth $11 million, were cast in the University of Arizona's own Mirror Laboratory. Innovatively, borosilicate glass was melted, weighted with aluminum blocks and then spun in a rotating furnace until a parabolic disc was formed. The mirror was then allowed to cool for several months, to avoid stresses building up, before it was removed and polished. It is estimated that this system, resulting in a lightweight "honeycomb" mirror that is 80 percent air, saved 56 tons (51 tonnes) of glass and up to two years of labor.

the **facts**

Between 1980 and 2005 the United States' production of wind energy increased from 10 to 6,740 megawatts.

In its 200-year lifespan, the farm will save the emission of more than 20 million tons (18 million tonnes) of CO_2.

The average US household consumes 1,000 kilowatt hours of electricity a month.

King Mountain Windfarm

HIGHS AND LOWS

For turbines such as those at King Mountain to operate efficiently, the average wind speed throughout the year must be at least 12mph (19kph); below this the generators automatically shut down. As wind speeds increase, generating power rises exponentially until the blades are rotating at their maximum rate of 17 rotations per minute in a stiff breeze of 35mph (56mph). Stronger winds than this bring no further increase in power and at 55mph (88mph) the automatic shut-off again comes into play, with the blades turned sideways to the wind. All these operations are performed automatically by using sensors mounted on the towers.

On the skyline of a mountain plateau, high above the oil fields of west Texas, lines of giant turbines rotate effortlessly in the desert wind. Texans are renowned for thinking big, and King Mountain Windfarm in Upton County is immense. Ranged along the mesa's southeastern and northwestern rims to take full advantage of prevailing winds, the 214 turbines sit on towers 197ft (60m) high with the tips of their rotors reaching 298ft (91m) up into the sky.

At the time of its completion in 2001 King Mountain was the largest windfarm in the world, and it is still one of the most powerful of its kind. With each of its Danish-designed turbines generating 1.3 megawatts of electricity, its full potential output of 278 megawatts is sufficient for the needs of 80,000 homes. What is more, the windfarm has revitalized the community it serves. With oil production falling every year, the fortunes of Upton County had long been in decline. Now, with the success of King Mountain stimulating the development of other windfarms in the area, the new industry contributes around $4 million in taxes every year for the benefit of schools and local projects. The small local town of McCamey proudly claims to be the "Wind Energy Capital of Texas."

Made to Measure

In many ways King Mountain is an ideal location for a windfarm. The 60sq-mile (150sq km) table-topped mesa rises almost 1,000ft (300m) above the surrounding land, and at 3,140ft (957m) above sea level, the turbines are exposed to almost constant wind. In addition, structures such as pumps and derricks are accepted as time-honored features of the landscape. Nonetheless, the extreme desert conditions meant that engineers had to modify some of the equipment. Cooling systems were improved to compensate for high daytime temperatures and, in order to prevent swirling dust from clogging the machinery, both the turbine nacelles (casings) and the steel towers were sealed and slightly pressurized, to prevent any particles from getting in. Also, since the air on King Mountain is 10 percent thinner than at sea-level, the gearing was adjusted to allow for slightly higher rotor speeds. Using standardized components and prefabricated towers, windfarms are relatively simple to construct and King Mountain was completed in just nine months, at a cost of $150 million.

TEXAS WIND RUSH

Texans showed little interest in windfarms until 1999, when a new state law required energy companies to invest in renewable resources. Encouraged further by generous tax breaks and positive responses from local residents, a huge construction boom then began in areas that promised suitable sites. By 2005 Texas was second only to California in terms of wind power. The boom has brought a new prosperity at a time when oil reserves are in decline, and with 700 megawatts now on line in Upton County alone, a major upgrade of the local transmission grid is now underway to cope with the demand.

There are 29,000 tons (26,000 tonnes) of concrete in the foundations of the arch and just under half that amount again packed between its steel walls.

The stainless steel outer wall weighs 977 tons (886 tonnes).

The two stairways to the top each have 1,076 steps.

Gateway Arch

America's tallest national monument to date is a gleaming arch of stainless steel that soars 630ft (192m) above the banks of the Mississippi River in St. Louis, Missouri. Monumental in its scale yet exquisitely delicate in its design, the arch forms the centerpiece of the Jefferson Memorial Park, which was established in 1935 to commemorate the westward expansion of America's frontier in the 19th century. Although it was always the intention to build some great memorial within the park, the project stalled during World War II and it was not until 1948 that a competition was announced for its design, with $50,000 offered as first prize. Many of the nation's leading architects took part, but the winner, Eero Saarinen, was a little-known designer of Finnish descent who produced an astonishing proposal that he described as "a landmark of our time." His arch owes nothing to the classical traditions that usually inform great monuments. Instead, it is a severely simple structure, with no ornament or detail, which stands as a colossal gateway to the vast open spaces of the West.

ROLLING UP AND DOWN

Since the curve of the arch made the use of conventional elevators impractical, a special transportation system was devised to reach the observation gallery. From the basement, trains of drum-shaped capsules rise within each leg, rotating through 155 degrees as they negotiate the curve on metal rails that are above the capsule as the climb begins and beneath it at journey's end. The trains, which each consist of eight five-seater capsules, take four minutes to reach the gallery, where there is space for 150 people, and a further three minutes to descend.

An Inverted Chain

Unfortunately Saarinen, who died in 1951 at the age of 41, never lived to see his project realized, but construction of the arch finally began in February 1963 and was completed in October 1965. Despite its delicate appearance, it is a structure of enormous strength that is designed to withstand tornados, swaying just 18in (46cm) in a wind of 150mph (240kph). The arch forms what is known as a catenary curve, the natural shape a chain assumes when hung between two points.

At ground level, the legs are set 630ft (192m) apart, exactly equal to the height of the arch. The legs are triangular in plan, with sides that taper in width from 54ft (17m) at the base to just 17ft (5m) at the top. Faced in stainless steel, they look like solid structures but are in fact elaborately engineered with double walls of steel around a hollow core. The gap between the walls is packed with concrete up to a height of 300ft (91m), with a lighter steel frame above. Within the hollow core, a specially designed transportation system takes visitors up from an underground museum to an observation platform at the apex of the arch that provides stupendous views to both the east and west of the city. The two stairways which lead to the top are reserved for maintenance workers and emergencies only.

MOVING WEST

The city of St. Louis developed after the Louisiana Purchase of 1803, when the United States paid $15 million to acquire more than 830,000sq miles (2.1 million sq km) of French territory to the west of the Mississippi River. The Jefferson National Expansion Memorial commemorates this momentous episode in American history, while its museum also celebrates the Lewis and Clark Expedition that set out from St. Louis in May 1804. Mapping unknown territory and negotiating with native tribes, the expedition reached the Pacific coast of Oregon in December 1805, laying the foundations of America's subsequent colonizing of the West.

High-speed elevators traveling at 1,600ft (488m) a minute take a million visitors a year up to the Skydeck on the 103rd floor.

The tower's total floor area is 105 acres (43ha).

The tower weighs 222,500 tons (200,000 tonnes), of which 76,000 tons (68,500 tonnes) are steel.

Sears Tower

Vast, uncompromising and bare of any ornamental flourishes, the Sears Tower in Chicago is not to everyone's taste. Completed in 1974, it represents the ultimate expression of the plain and rectilinear International Style that for many years remained the norm for high-rise office blocks in almost every major city in the world.

The Chicago giant is, however, of a quite different order to the run-of-the mill monoliths associated with the style. More than 30 years after its construction, the Sears Tower is only now recognized as a masterpiece of modernist design. Its sheer curtain walls, which are clad in black anodized aluminum and bronze-tinted glass, are remarkable, not only for their scale, but also for their purity and elegance of form. Completed at a cost of $175 million—a relatively modest sum in relation to its size—it was also a ground-breaking building in terms of its structural design.

The brief that Sears Roebuck gave to the architectural practice Skidmore Owings & Merrill (SOM) was to design a new company headquarters that would be the largest and the tallest office block in North America. It was the engineer, Fazlur Khan, who came up with idea of constructing the tower from a square of nine self-supporting steel tubes, or "mega-modules." It was the architect, Bruce Graham, who decided that the tubes should terminate at different heights to give the tower its distinctive stepped silhouette.

Stepping Up

For the first 50 of the building's 110 stories, the tower rises as a uniform rectangular block, with floors of 50,000sq ft (4,650sq m). Thereafter it assumes an ever-changing shape, with seven tubes continuing up to the 65th floor, five to the 89th and three rising to the tower's full height of 1,450ft (442m). Until the completion of the Petronas Towers in Malaysia in 1997, it was the world's tallest skyscraper, a position it reclaimed in 2000 when a new antenna increased its height to 1,736ft (529m).

The tower remains the world's largest office block, with a floor area second only to the Pentagon. Its scale can be gauged from some staggering statistics. There are 25 miles (40km) of plumbing, 1,500 miles (2,400km) of wiring, and 104 elevator cars.

SEARS ROEBUCK

From modest beginnings selling watches in 1886, Sears Roebuck eventually became the largest retail business in the world, with a workforce of more than 300,000 employees. Throughout much of the 20th century the company's vast mail-order catalog, sometimes called "the consumer's bible," helped to define the American way of life. However, by the time the Sears Tower was completed in 1974, the retail world was changing and the company was already in decline.

In 1995 the tower was sold and it is now occupied by more than 100 different businesses. Finally, in 2005, Sears Roebuck was taken over by the US retail giant Kmart in a deal worth a staggering $11 billion.

CHICAGO ARCHITECTURE

Following a disastrous fire in 1871 that destroyed more than 17,000 buildings, Chicago rose from the ashes as a city pioneering modern forms of construction and design. The Home Insurance Building of 1885 saw the first use of a steel frame in place of masonry walls, and in 1892 the 22-story Masonic Temple became, albeit briefly, the tallest building in the world. In later years, architects such as Mies van der Rohe and Frank Lloyd Wright maintained the city's reputation for cutting-edge design, while since the 1960s Skidmore Owings & Merrill have built some of the world's tallest towers.

The fast-track building program aimed to complete two floors of the tower a week.

There are 202 rooms and 8 suites in the hotel.

Dampers are used in all sorts of buildings, including stadiums where the rhythms of crowds walking can cause vibrations, swaying, and twisting.

Park Tower

LUCIEN LAGRANGE & ASSOCIATES

During his early career French architect Lucien Lagrange worked for the high-profile practice Skidmore Owings & Merrill. He established his own practice in Chicago in 1985. His firm is known for its elegant work—ranging in style from beaux-arts and art nouveau to high modernism—and embracing luxury new-build hotels and apartments.

The firm also specializes in interior and retail design, along with the refurbishing and remodeling of older buildings such as Chicago's Union Station, the Insurance Exchange, and Blackstone Hotel.

The 844ft (257m) tower rises up in steps from stores at street level. The first 20 floors are occupied by the luxury Park Hyatt Hotel. Contemporary, fashionable and understated, the lobby has as its focal point German pop artist Gerhard Richter's (b. 1932) painting *The Piazza del Duomo-Milan*. Above the hotel are 45 floors of luxurious apartments.

Keeping Steady

On completion, Park Tower was one of the world's tallest buildings to be clad with architectural pre-cast concrete. Chosen in preference to stone on the grounds of cost and ease of construction, the concrete panels were packed into racks on a trailer and lifted one by one into position, where they were bolted and welded. The 3,152 panels cover a surface area of around 2,370sq ft (2,220sq m) and weigh a total of 8,750 tons (7,940 tonnes).

Chicago is not known as the Windy City for nothing, and to counterbalance the effects of high winds, the tower has a tuned mass damper (TMD) inside the copper-clad pyramid that crowns the building. The damper is a massive steel pendulum, weighing 300 tons (272 tonnes), which is suspended from four cables inside a square cage. TMDs are so named because they move "in tune" to the movements of the building to maintain stability.

With all the glamor of Chicago's greatest skyscrapers, the 67-story Park Tower in the heart of town on North Michigan Avenue—known as the Magnificent Mile—is a distinguished addition to the cityscape. It has a powerful art deco look with its creamy-beige decorative cladding, rounded corners, curved balconies and a copper-clad, pyramid-shaped crown. Designed by Lucien Lagrange & Associates, the tower was commissioned by the Hyatt Development Corporation and completed in 2000 at a cost of $92 million.

CHICAGO
WATER TOWER

Constructed in 1869 using limestone blocks from Illinois, the 154ft (47m) Chicago Water Tower was designed by William W. Boyington (1880–98). It houses a 131ft (40m) standpipe, which was needed to equalize the pressure of the water pumped from the pumping station to the east. In May 1969, the year of its centennial anniversary, the Chicago Water Tower was selected hy the American Water Works Association to be the first American Water Landmark. It now houses a visitor information center and has become one of Chicago's best-known features.

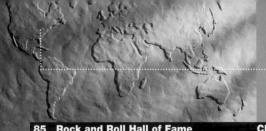

- The museum opened in September 1995.
- The idea for the museum came from the founder of Atlantic Records, Ahmet Ertegun.
- Chuck Berry, Bob Dylan, Aretha Franklin, Johnny Cash, The Pretenders, and Bruce Springsteen all played at the inaugural concert.

Rock and Roll Hall of Fame

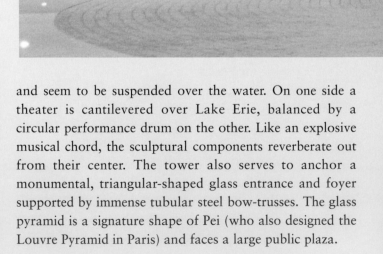

Bursting with the exuberance and dynamism of its subject matter, the Rock and Roll Hall of Fame and Museum in Cleveland, Ohio, is designed as a sculptural explosion of three-dimensional geometric shapes.

Sitting on the shores of Lake Eyrie between the Cleveland Browns Stadium and Burke Lakefront Airport in downtown Cleveland, the museum was designed by I. M. Pei and opened in September 1995. With such priceless exhibits as Janice Joplin's multicolored psychedelic Porsche, Bruce Springsteen's outfit from the cover of his album *Born in the USA*, Bono's first guitar, and the Sergeant Pepper uniforms worn by John Lennon and Ringo Starr, this museum is the world's first dedicated to the living heritage of rock and roll.

Seen from the water, the museum has at its heart a 165ft (50m) tower that rises from the harbor edge. Two bold, solid geometric shapes explode sideways from the tower and seem to be suspended over the water. On one side a theater is cantilevered over Lake Erie, balanced by a circular performance drum on the other. Like an explosive musical chord, the sculptural components reverberate out from their center. The tower also serves to anchor a monumental, triangular-shaped glass entrance and foyer supported by immense tubular steel bow-trusses. The glass pyramid is a signature shape of Pei (who also designed the Louvre Pyramid in Paris) and faces a large public plaza.

GETTING IN

In addition to supporting the development of the museum, the Rock and Roll Hall of Fame Foundation organizes the annual nomination, election, and induction of new members into the Hall of Fame. The nominees, chosen by a committee of historians and musicologists, are eligible if they have released a record at least 25 years prior to induction. They are then voted for by an international group of music industry professionals, including producers, broadcasters, journalists, and performers. Since 1986, approximately 180 stars have joined the pantheon of rock greats.

A Musicians' Monument

The museum interiors range from dramatic, multistory, open-plan spaces to more intimate exhibition areas. The material palette is simple: "white" architectural concrete at the base walls of the building, off-white aluminum panels with deep reveals in a contrasting gray for the skin of the tower, gray window mullions and framing for the clear glass pyramid, and charcoal-gray flooring. The pale color of the tower provides a dramatic sculptural background to

CITY FOUNDING

General Moses Cleaveland founded the City of Cleveland in 1796. The name changed to its current spelling in 1831 when the letter "a" was dropped in order to fit the city's name onto a newspaper masthead.

Originally a frontier village, Cleveland grew into a manufacturing and business center for Northern Ohio. Today the city is the headquarters of both manufacturing and service industries, as well as a growing destination for tourism and conventions.

Cleveland is the 16th largest metro area and the 15th largest consumer market in the United States. Nearly 500,000 people live in the City of Cleveland, making it the 30th largest city in the country.

the activity within the glass foyer, especially when illuminated at night. Unusually, the bulk of the museum has been constructed underground in order to create a suitably controlled environment for the museum's many hands-on interactive installations.

- The six trains each have five cars seating 18 people each, permitting up to 1,500 rides per hour.
- It was the world's tallest rollercoaster until 2005, when it was surpassed by a ride at Six Flags, New Jersey.
- The ride never runs during rain, as raindrops are very painful when they strike the face at 120mph (192kph).

Top Thrill Dragster

When the Cedar Point theme park in Ohio decided to create a new attraction to add to the 16 rides already on offer, they turned to the Swiss engineering firm of Intamin AG, which specializes in extreme fairground-ride design. Their brief was very simple: this was to be the tallest, fastest, and most terrifying rollercoaster in the world. And when the Top Thrill Dragster opened to the public in 2003, the $25 million ride was a triumph in all three respects.

HISTORIC COASTERS

In the 18th century the Russian aristocracy enjoyed the thrill of artificial sledge-rides, using ice-coated ramps descending from towers up to 80ft (25m) high. By 1817 there were rides in France that featured carriages on looping rails, but it was not until 1884 that the world's first true rollercoaster, or "switchback railway," opened on Coney Island in New York.

Since then American theme parks have competed to provide ever more extreme examples of the form, with a variety of high-tech launching systems now leading to rides offering unprecedented heights, speeds and excitement.

The world's oldest surviving coaster, at Lakehurst Park in Pennsylvania, was built in 1902 with a height of 48ft (15m) and a top speed of 10mph (16kph).

Fastened to a soaring skeleton of steel tubes, the twisting, looping track rises to an astonishing 420ft (128m), the height of a 40-story skyscraper. Within just four seconds of the start cars are hurtling down the straight at 120mph (192kph), propelled by the colossal power of a hydraulic launching system. Then, once the car has climbed the tower and teetered briefly on the crest, there is the sheer, screaming terror of a free-fall descent with the car spinning on its axis as it plummets at 175ft (53m) per second.

Licensed to Thrill

Despite its madcap image, the giant rollercoaster is a highly complex structure that took more than three years to design. At one stage Intamin built a scale model in a field near their headquarters to test whether the launch system had sufficient power to propel cars up the tower. The challenge the designers faced was to build a ride that, while inducing a near-death experience, was actually completely safe. The braking system, consisting of copper fins that run between magnets in the track, will operate even if all the electrics fail, while the lap restraints are hydraulically controlled and will only open when a ride is over.

The structure is immensely strong, with the 2,800ft (854m) track supported on tubular steel weighing 1,150 tons (1,043 tonnes), with 151 support columns resting on solid concrete foundations. Every twist and turn along its length, including the sheer drop from its tower, has been carefully designed using sophisticated computer programs, to ensure that G-forces remain within safe limits.

HYDRAULIC LAUNCH

In the past, rollercoaster trains were hauled up to the highest point on chains before release, but the Top Thrill Dragster employs a far more spectacular method of launch. A system of hydraulic pumps, motors, and accumulators catapults the trains along the initial straight, transmitting up to 10,000 horsepower during the first six seconds of the ride. With each launch, computers calculate the precise amount of power required to propel the train up to the highest point, but adverse winds or excess weight can sometimes mean that trains fail to make the grade and have to roll back to the start for a second attempt.

The bridge cost $37 million to build.

To reduce the need for rivets, the prefabricated sections were precision-made to tolerances of 100th of an inch (0.25mm).

The bridge replaced a 40-minute drive into the valley along hairpin mountain roads.

New River Bridge

The New River Bridge is an engineering masterpiece that springs across an awe-inspiring gorge in the hills of West Virginia. At the time of its completion in 1977, it was the largest steel arch bridge that had ever been built, with a clear central span of 1,700ft (518m) crossing a steep-sided valley 876ft (267m) deep. It was the nature of the site that dictated the design; steel is relatively light yet immensely strong, while an arch, owing to its geometric form, can support enormous downward stresses, which are transferred into the ground on either side. Another factor that the design engineers Michael Baker Jnr. Inc. had to take into account was the environmental impact on the area. However, this is minimal, as the bridge appears to soar lightly through the sky, as though flung out from the valley sides.

SALT AND STEEL

For many years the road across the bridge was kept ice-free in winter by using salt and grit. But salt is an enemy to steel, and after several years it was discovered that structural supports were showing signs of dangerous corrosion. In 1998 a team of specialists had to be employed, at a cost of $3 million, to undertake a program of cleaning and repair, blasting off the rust with ultra-powerful pressurized water-jets. Having learnt from this mistake, the Department of Highways now uses a special salt-free solution to keep the bridge clear of ice and snow.

Building from Above

Before construction could begin in the summer of 1974, extensive preparations had to be completed. In the valley's sides, abandoned mineshafts were filled with gravel to provide support for the massive concrete bases of the arch. Also, since the bridge has a total length of 3,030ft (924m), approach spans were built out from each side. Then the engineers faced the challenge of assembling the enormous arch. For this, two pairs of temporary towers 330ft (100m) tall were built and, with the aid of a helicopter, a cableway was strung across the gorge. Prefabricated sections of the arch could then be hauled from the banks on trolleys, lowered from above and riveted together. As the sections arched ever farther over empty space, they were prevented from collapsing by cables fastened to the towers until the final section, the vital "key" of the arch, could be dropped precisely into place. Lastly, with the arch providing a strong backbone of support, the road deck was built out on a series of steel struts up to 305ft (93m) high. The bridge was finally completed in October 1977.

BRIDGE DAY

On the third Saturday of October every year the bridge is closed to traffic for a day devoted to risk and fun. Since the first "Bridge Day" was held in 1980, what started as a local festival has grown into the world's largest gathering of extreme-sports enthusiasts, attracting crowds of up to 200,000 people and featuring a program of spectacular events. BASE jumpers can experience the thrill of freefalling for up to nine seconds before opening their parachutes, teams of rapelers compete to make the fastest climb, while whitewater rafters negotiate the rapids far below the New River Bridge.

Construction of the bridge began in 1998 and traffic lanes opened in stages between 2003 and 2005.

The 116 cable stays are made from 1,820 miles (2,900km) of steel wire.

The bridge is named in honor of Lenny Zakim (1953–99), a civil rights leader who bridged divides.

88 Zakim Bridge **Boston,** USA

CHRISTIAN MENN

While teams of experts contribute to any major engineering project, the appearance of the bridge reflects the technical and artistic genius of its architect, Christian Menn. Born in Switzerland in 1927, Menn became a bridge-designer in the 1950s, when he pioneered the use of pre-stressed concrete spans as an alternative to arched construction.

In later years he became increasingly intrigued by the aesthetic properties of engineering, which he saw as inseparable from practical considerations. In searching for a balance between beauty, efficiency, and cost, he has been responsible for a number of astonishing structures, including the curvaceous Ganter Bridge across the Simplon Pass, in Switzerland, in 1980.

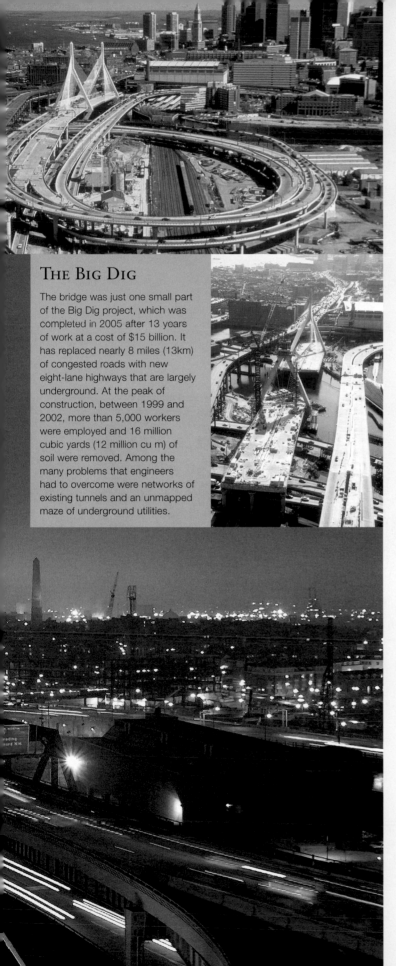

Zakim Bridge

T|he greatest engineering challenges often stimulate the most ingenious designs. This was the case when it came to replacing the two-level bridge that had carried Boston's traffic across the Charles River since 1959. As a vital link in the city's ambitious Central Artery/Tunnel Project (the Big Dig), the new bridge would have to be capable of carrying up to 200,000 vehicles a day. Furthermore, the site itself presented seemingly impossible constraints, with subway tunnels just beneath the surface and a ventilation building that could not be moved, pressing close to the side. Also, to avoid traffic chaos, the existing bridge could not be demolished until the new one was complete. Given such conditions, it is not surprising that the Leonard P. Zakim Bunker Hill Bridge (or Zakim Bridge) is a one-off.

Asymmetrical Design

Carrying 10 lanes of traffic and with a total length of 1,457ft (444m), the bridge is built on an impressive scale. Measuring 185ft (57m) across, it is the widest cable-stayed bridge in the world. It is also the world's only asymmetric bridge, with one side of the road deck cantilevered out beyond the cable-stays and towers to carry traffic heading downtown. In addition, the 330ft (100m) tower on the north bank of the river is 27ft (8m) taller than its southern twin, although both rise 273ft (83m) above the road. The explanation for this anomaly is that the whole bridge slopes at 5 degrees to link an elevated highway with a tunnel. Indeed, almost every oddity about the bridge turns out to be an ingenious solution to a problem. The bases of the towers bend sharply inwards at an angle of 55 degrees underneath the road-deck to avoid nearby structures and to straddle subway lines below. The unusual pattern of the cable-stays is also due to site restrictions; while the 745ft (227m) main span is hung from sets of cables fastened to its sides, the bridge approaches hang from sail-like fans fastened to the central reservation. Such details all contribute to the bridge's distinctive form.

THE BIG DIG

The bridge was just one small part of the Big Dig project, which was completed in 2005 after 13 years of work at a cost of $15 billion. It has replaced nearly 8 miles (13km) of congested roads with new eight-lane highways that are largely underground. At the peak of construction, between 1999 and 2002, more than 5,000 workers were employed and 16 million cubic yards (12 million cu m) of soil were removed. Among the many problems that engineers had to overcome were networks of existing tunnels and an unmapped maze of underground utilities.

The tower is 914ft (279m) high and has 59 stories.

The main shaft of the tower is raised on columns 114ft (35m) high.

It took five years and $40 million to acquire all of the leases of the buildings on the site before construction could begin.

the facts

89   Citigroup Center   New York, USA

Citigroup Center

With its rakish, sloping top and base raised on stilts, the 914ft (279m)-high Citigroup (formerly Citicorp) Center is a powerful corporate icon and one of the most distinctive towers of midtown Manhattan. The scheme evolved in the early 1970s when Citigroup (then called the First National City Bank) was looking to expand in the neighborhood at the same that St Peter's Lutheran Church, occupying a prime site on Lexington Avenue, was in need of extra finance. A deal was struck and Citigroup bought the land on the condition that a new church would be built on the site.

Raised on Stilts

To accommodate the new St Peter's, the tower was raised on a central supporting core with four columns placed in the middle of each face of the building—not, more conventionally, at the corners. This open space was then used to create a sunken plaza along with a shopping center, restaurants and cafés, a seven-story atrium, and a refurbished subway station, with the church in the northeast corner. The scheme prompted a revival of interest in the area and it has subsequently gone through a period of redevelopment and revitalization.

Distinguished American engineer William LeMessurier (b. 1926) conceived the innovative structural design in the form of a chevron system of diagonally placed steel girders. Although barely visible from the outside as the tower is clad in glass and aluminum panels, they are a striking presence in the interior. The 164ft (50m)-high sloping top section was originally conceived to incorporate penthouse apartments, but planning permission for this was refused and the space now houses the mechanical equipment required by the building. This includes a massive tuned mass damper, or stabilizer—a 400-ton (360 tonne), 30ft (9m)-square block of concrete used to counterbalance the effects of winds on the building.

The building incorporates many environmental features which were cutting edge in their time—double glazing, heavy insulation, heat recovery from lighting and electrical equipment, energy-efficient lighting and the effective use of daylight for office areas. To this day, such is the efficiency of these systems that the building requires no heating until the outside temperature falls below freezing.

ST PETER'S LUTHERAN CHURCH

Designed in an abstract, octagonal form, like a pair of hands clasped in prayer, the new St Peter's Lutheran Church is the work of architect Hugh Stubbins (b. 1912), who also designed the tower. Its angular, skyward-pointing shape wrapped in gray granite incorporates tall vertical windows and sculpture by Russian-born American abstract expressionist sculptor Louise Nevelson (1899–1988). St Peter's has long been known as the jazz church. The former building was the venue for legendary jazz musician Louis Armstrong's (1901-71) memorial service, and for many years both old and new churches have offered free jazz performances.

RUNNING REPAIRS

In 1978, just one year after the skyscraper's completion, the structure required urgent remedial works. It was discovered that changes made during construction replaced expensive welded joints with less costly bolted ones. Concern was raised that the tower might be unable to withstand high-level winds.

The solution proposed by William LeMessurier was to hire a crew of welders who would remove internal sections of floor and wall to expose the 200 bolted joints and then permanently weld steel plates over each one to correct the problem. Amazingly, the work was undertaken out of office hours during a period of three months without disrupting the occupants.

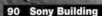

The building is 647ft (197m) to the top of the pediment.

The vertical strips of windows have been likened to the radiator grille of a Rolls Royce car.

The controversial decorative top of the tower alludes to furniture designs by the English cabinetmaker Thomas Chippendale (1718-89).

Sony Building

POSTMODERNISM

The phrase postmodernism first came into use in the mid-1970s and its meaning continues to be a matter for debate. Outside of architecture, the word has been used to describe a theoretical point of view that questions the status quo and widely accepted beliefs. When applied to architecture, it has come to describe an antithesis to the purity and rationalism of modernism. Often with an edge of irony, postmodernist architecture embraces ornament and color and frequently borrows elements from the Classical tradition. Architects who are closely associated with postmodernism include Americans Michael Graves (b. 1934) and Robert Venturi (b. 1925).

Generally hailed as the first postmodern skyscraper, the former AT&T Building in New York marked a break with the boxy, glass-clad, corporate architecture of the previous quarter century. Simultaneously pointing to the future yet capturing the romance of the skyscrapers of the 1920s and 1930s, the project was described by its architect Philip Johnson as the job of his life.

The slender, 34-story building occupies a constrained site on Madison Avenue at 56th Street, in Manhattan. The main features of the base, which encloses a huge public plaza, are the monumental entranceways. The tallest of these is an arched portal measuring some 110ft (34m), flanked on each side by three 60ft (18m)-high rectangular entrances. The main shaft of the tower rises with its windows positioned in nine vertical stripes and at the top is the distinctive broken-pediment, behind which lies the building's mechanical service equipment. The building is clad in a pink-gray marble from the same quarry that provided the stone for New York's well-known train station, Grand Central Station.

A Confusion of Styles

When completed in 1984 the tower provoked comment for a number of reasons. Firstly, it broke away from the pure lines of modernist, all-glass skyscrapers that were so prevalent at the time, and secondly it was designed by the controversial figure Philip Johnson who had first promoted modernism and was now apparently turning his back on his beliefs. Critics perceived the building as a confused pastiche of styles including Gothic and art deco. Johnson was robust in his defense of his work, stating that he believed he was leaving a building which would be regarded in a positive light by future generations. However, along with its detractors, the building did have its admirers—the critic of the *Chicago Tribune*, Paul Gapp, wrote that the building had brought postmodernism into the corporate mainstream.

Philip Johnson has designed numerous other high-rise projects other than the Sony Building. He joined forces with partner John Burgee and from 1967 through 1987, the company's output was prodigious. Amongst other buildings, they designed International Place in Boston, Pittsburgh Plate Glass Building in Pittsburgh, the Crystal Cathedral in Los Angeles, the Dade County Cultural Center in Miami, and outside the US, the National Center for Performing Arts in Mumbai, India.

PHILIP JOHNSON

A controversial and potent force in 20th-century architecture, Philip Johnson (1906–2005) was a critic, author, historian, and museum director all before designing his first building at the age of only 36. In the late 1940s he designed a home for himself in Connecticut known as the Glass House—a startling glass box that has become an icon of modernism. Johnson organized the first visit to the United States of European architects Ludwig Mies van der Rohe (1886–1969) and Le Corbusier (1887–1965). In the postwar years he revised his views on architecture, culminating in this building, his most controversial.

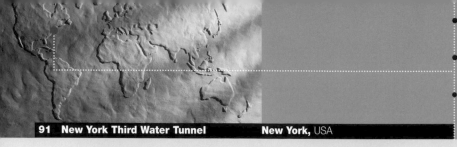

The largest valve chamber, 250ft (76m) below a park in the Bronx, measures 620ft by 43ft (189m by 13m).

New York consumes 1.1 billion gallons of water a day, equivalent to 136 gallons (618 liters) per resident.

For each mile of tunnel, 88,000 cubic yards (67,000cu m) of rock and soil has to be removed.

New York Third Water Tunnel

Hundreds of feet below the sidewalks of Manhattan, Brooklyn, and the Bronx, underground construction crews have been at work for more than 35 years, advancing one of the longest tunnels in the world at a rate of up to 17 yards (16m) a day. Although it has never been accorded any fanfare of publicity, New York's Third Water Tunnel is one of the biggest civil engineering projects ever undertaken in America. By the time it is completed in 2020, it will have taken half a century to build, cost $6 billion, and extend for 60 miles (97km)—almost twice the length of the Channel Tunnel between England and France.

The need for a new tunnel was recognized as early as the 1950s, when a fundamental flaw became apparent in the way that the existing system had been designed. The two great tunnels that supplied nine million people with their water every day were showing signs of wear and tear. The problem lay in the fact that the tunnels could not be shut down or diverted to allow for inspection or repair.

Mechanical Moles

After many years of careful planning, work on the massive project finally began in 1970 when tunnel-workers, known as "sandhogs," started to drill and blast the first 13-mile (21km) section from Hillview Reservoir in Yonkers to Fifth Avenue. Working at a depth of up to 800ft (244m), the crews endured hideously challenging conditions and, despite every modern safeguard, 24 lives had been lost by the time the section was complete in 1994. Since then both safety and the rate of progress have been greatly improved by the use of a tunnel boring machine, a mechanical mole with a rotating head of steel teeth that bores a 24ft (7m) diameter tunnel without the need for hand-drills or explosives.

The new tunnel will eventually extend far beyond the limits of New York City to Kensica Reservoir in Westchester County, but the most important innovation lies in the introduction of four great valve chambers that link it to the older system. In these, the water-flow is diverted along conduits that can be individually controlled, allowing sections of the gravity-fed network to be closed off for repair without any disruption to water supply, essential to avoid disruption to a city as vast as New York.

PURE AND SIMPLE

New Yorkers are justly proud of the natural purity of their tap water, and although a controversial new filtration plant will shortly be provided for the Croton system, supplies from the Catskills will remain as unadulterated as a mountain spring. Much of the watershed, which covers an area of 1,600sq miles (4,142sq km), is in private ownership, and over the past 30 years many innovative agreements have been signed to confirm the proper management of its streams and forests to ensure there is no pollution of the water.

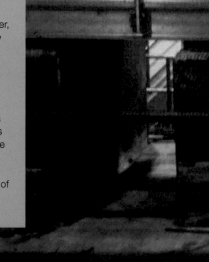

A THIRSTY CITY

New York's first city reservoir was constructed in 1776 on the east side of Broadway, using hollow logs to pipe water to households. In 1842 the city built the Old Croton Aqueduct, providing 90 million gallons (405 million liters) of water a day from the Croton River in Westchester County, where a second reservoir was opened in 1890. The much larger Catskill and Delaware system was developed by the Board of Water Supply throughout the first half of the 20th century, and now accounts for 90 percent of supply. Overall, there are 19 reservoirs with a storage capacity of 580 billion gallons.

The tower's height of 1,776ft (541m) symbolizes 1776 when the US issued its Declaration of Independence.

The mast is 276ft (84m) tall.

As a memorial to those who died in September 2001, the "footprints" of the Twin Towers will be transformed into water-filled pools called Reflecting Absence.

Freedom Tower

n the wake of the horrifying events of September 11, 2001, which saw the Twin Towers of the New York World Trade Center destroyed in a terrorist plane attack, there ensued a long debate about how and what to build on the site. A number of new designs emerged, but eventually the Freedom Tower, by David Childs of Skidmore Owings & Merrill, was chosen to become the landmark tower on the site. The Freedom Tower will eventually be joined by a further four towers, a cultural center, a performing arts center, a transportation hub, and memorial park—all part of a masterplan by the architect Daniel Libeskind.

Looking to the Future

The tower is conceived as a gleaming, crystalline skyscraper in the shape of an obelisk with chamfered sides, complete with an observation deck and mast. The idea behind the tower is that it should represent optimism and the future, while, at ground level, the memorial will look to past events and remembrance.

The "footprint" of 200ft (61m) square is the same as each of the former Twin Towers. At street level, the building rises from the plaza to an 80ft (24m)-high public lobby topped by a series of service floors to form a 200ft (61m)-high building base. From here, 69 floors of offices rise to 1,120ft (341m). Service floors, two floors for the Metropolitan Television Alliance and restaurants culminate in an observation deck and glass parapet at 1,362ft (415m) and 1,368ft (417m) respectively—the same heights as the original towers. The mast rises to 1,776ft (541m).

Safety measures include a blast-proof, shimmering stainless-steel and titanium cladding for the base, biological and chemical filters for the air supply system, extra wide escape stairs, and high levels of fireproofing. All the building's safety systems, including communications, sprinklers, stairs and elevators, are encased inside a 3ft (1m)-thick core wall. Environmentally aware features include the recycling of rainwater for cooling the building and the use of sustainable timber and energy-saving glass. Even construction vehicles will use ultra-low sulphur diesel fuels. The building is expected to be complete by 2010.

SKIDMORE OWINGS & MERRILL

A large architecture practice, SOM was founded in 1936 in Chicago by Louis Skidmore and Nathaniel Owings. It took its current name three years later when John Merrill joined the office. The company, now with offices across the world, has a long history of creating landmark corporate buildings—often on an impressive scale—and incorporating innovative engineering solutions. Among its first designs to draw praise was Lever House in New York, completed in 1952, a slick curtain-walled office block in the International style. Since then, its best-known projects include the John Hancock Center and the Sears Tower, both in Chicago.

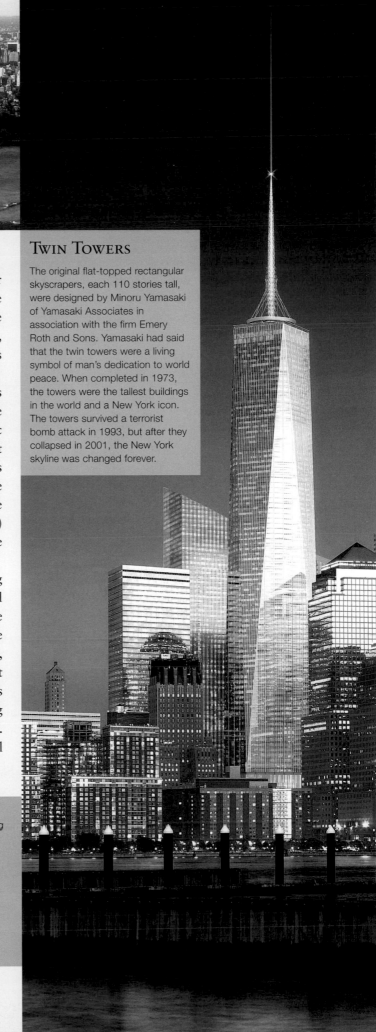

TWIN TOWERS

The original flat-topped rectangular skyscrapers, each 110 stories tall, were designed by Minoru Yamasaki of Yamasaki Associates in association with the firm Emery Roth and Sons. Yamasaki had said that the twin towers were a living symbol of man's dedication to world peace. When completed in 1973, the towers were the tallest buildings in the world and a New York icon. The towers survived a terrorist bomb attack in 1993, but after they collapsed in 2001, the New York skyline was changed forever.

The surface area of the sphere is clad in 2,474 aluminum panels, giving 5,599,663 acoustic-enhancing perforations.

The circumference of the sphere is 273ft (83m) and the diameter is 87ft (27m).

The cube can accommodate 1,200 people.

Rose Center

The American Museum of Natural History in New York contains the Rose Center for Earth and Space, which itself contains the show-stopping Hayden Planetarium. On the north side of the vast museum complex, this unmissable contemporary landmark consists of a metal-clad sphere apparently suspended and floating in a huge glass cube.

Designed by the New York-based firm Polshek Partnership Architects, the $210 million Rose Center, which opened in February 2000, is an exhibition, research, and education facility. Along with the rebuilt Hayden Planetarium, it comprises the Cullman Hall of the Universe, explaining the discoveries of modern astrophysics, the Gottesman Hall of Planet Earth, dedicated to exploring how the Earth works, and numerous additional visitor facilities and gardens.

The Hayden Planetarium

The showpiece Hayden Planetarium is composed of two main elements—the glass cube and the metal sphere. At the time of its completion, the cube was clad in the largest glass curtain wall in the United States. Its near-invisible structure is constructed from a system of tubular steel wall trusses braced by high-strength stainless-steel tension trusses. This futuristic form reflects the topic–astronomy–in which the Planetarium specializes.

Inside this ethereal cabinet is the large sphere which houses the 432-seat Space Theater, where visitors are taken on a virtual exploration of the universe with changing commentaries by well-known actors such as Harrison Ford and Tom Hanks. This virtual reality simulator is the largest and most powerful in the world, and the star projector is the most advanced in the world. In the lower half of the sphere is the 160-person Big Bang Theater, where a movie shows the explosive beginnings of time and space. Visitors stand, looking down onto the circular screen, and are entertained by surround sound, and laser and lighting effects, and educated about the origins of the universe.

POLSHEK PARTNERSHIP

Polshek Partnership Architects is a 150-person firm known for architectural excellence and its longstanding commitment to cultural, educational, governmental, and scientific institutions. It has been in New York since the practice was founded by James Polshek in 1963. As well as the Rose Center, the practice has designed the Santa Fe Opera Theater in New Mexico, the Carnegie Hall renovation in New York, and the masterplan for the renovation and expansion of the Brooklyn Museum of Art, including redesigning its historic facade.

THE COLLECTION

Founded in 1869, the American Museum of Natural History, on the west side of Central Park in Manhattan, has an astonishing collection extending to almost 40 million specimens from the natural world. It is housed in a range of buildings that extend for four city blocks and cover 20 acres (8ha). Altogether there are 42 permanent exhibition halls dedicated to every aspect of natural history, from reptiles and amphibians to birds and human biology.

Highlights of any visit include the two halls of dinosaurs, the hall of meteorites, including one 34-ton (31 tonne) chunk of meteorite and three Moon rocks, and the hall of minerals and gems. Here you'll find the Star of India sapphire, along with other extraordinary exhibits such as a 39ft (12m) model of a giant squid found in Newfoundland.

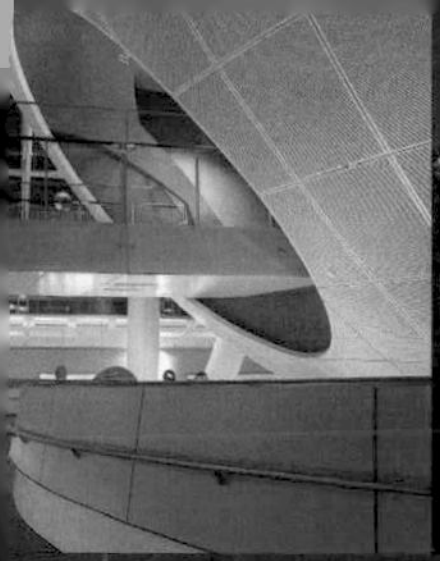
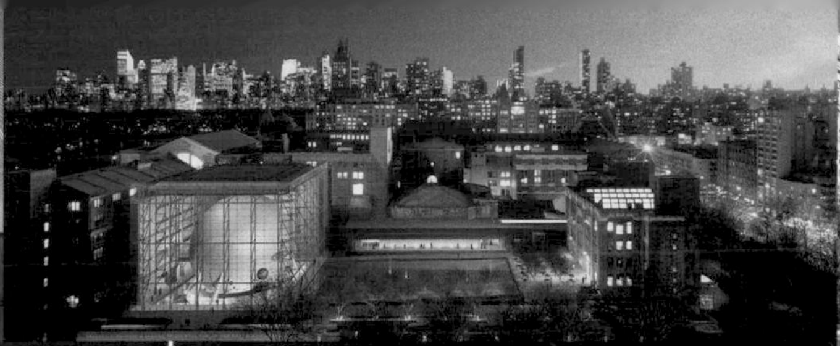
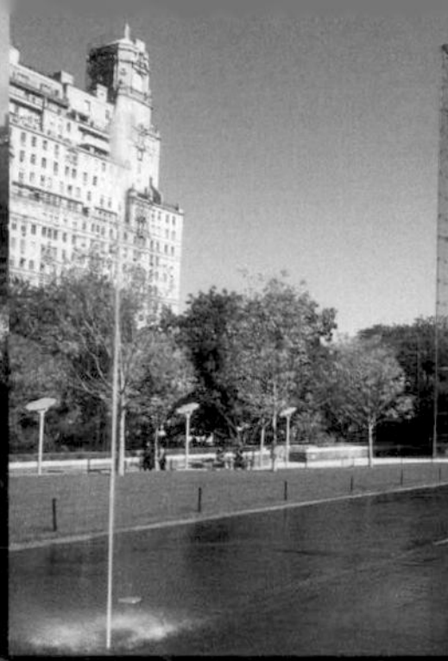
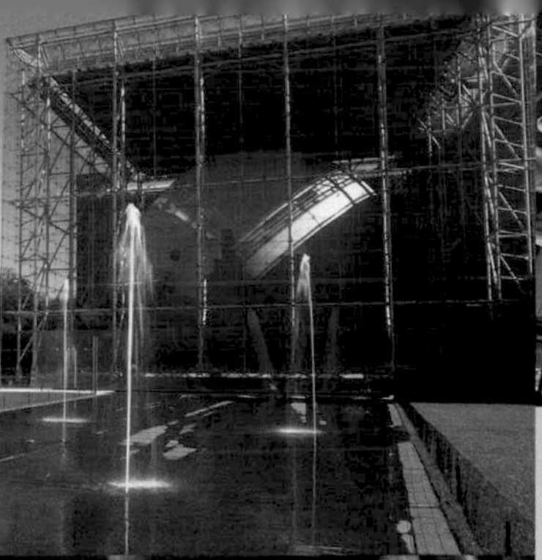
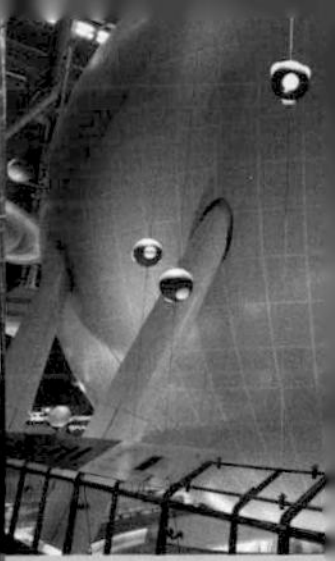

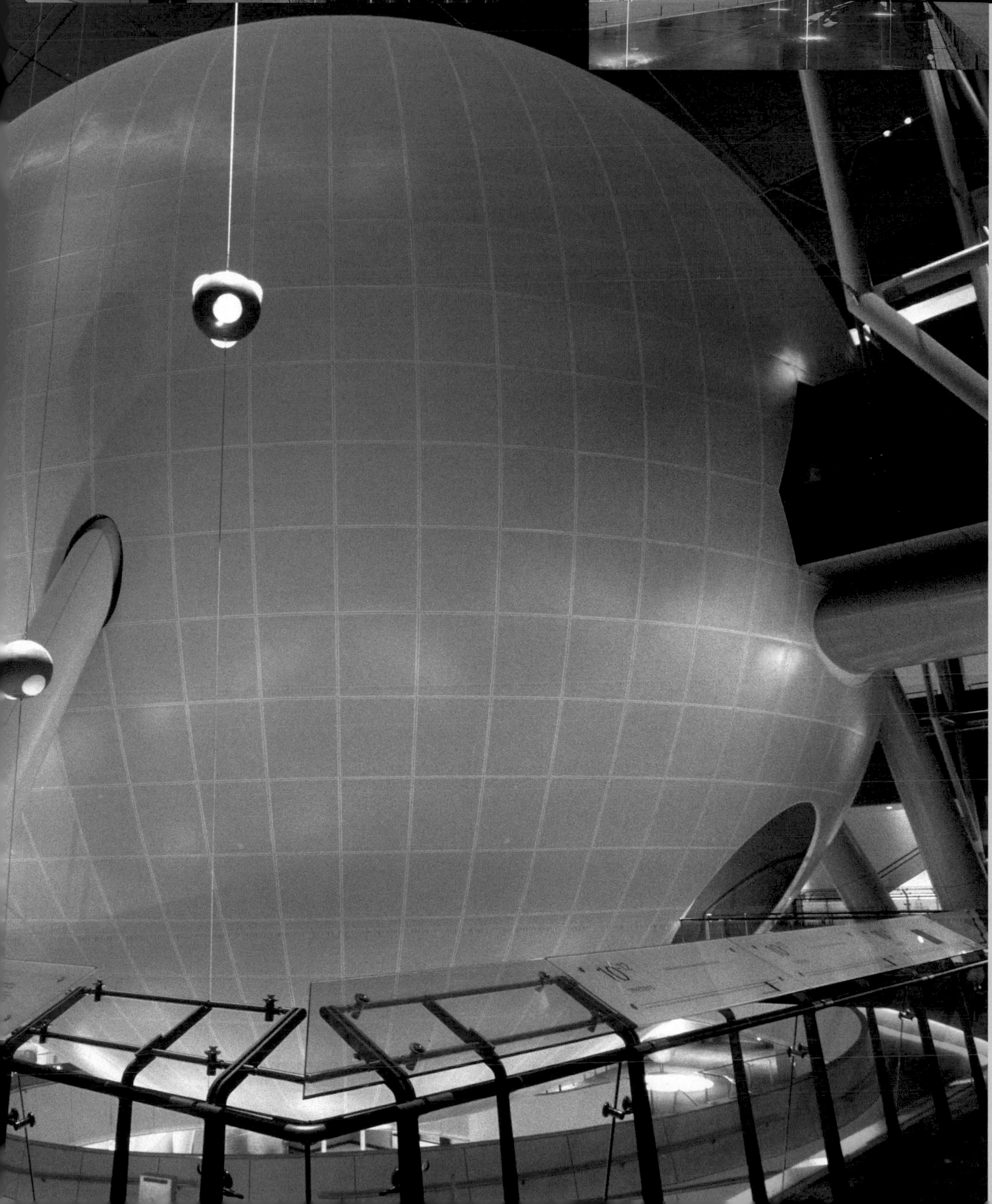

- The hotel exterior is wrapped in 8,000 sheets of glass, backpainted in different colors.
- Standing 532ft (162m) tall, this tower was the largest new-build hotel to open in Manhattan in 17 years.
- There are more than 40 theaters within a few minutes' walk of the hotel.

Westin Hotel

Among the most exuberant new towers to arrive on the New York skyline is the Westin New York Hotel on Times Square. Standing 45 stories and incorporating 863 rooms, the distinctive building is as flamboyant as its location, soaring above the traffic and colorful billboards of Times Square in the Broadway theater district of Manhattan.

Between Heaven and Earth

The design is conceived as a symbolic fusion of earth and sky spliced by a curving beam of light. Finishing in a skyward-pointing flourish, the western "sky" half of the tower is sheathed in vertically striped glass in intense steel-blue and pale purple. The eastern portion sits a few stories lower than its partner. It visually anchors the building to the ground by the horizontal stripes of shimmering bronze and copper glass that depict the earth. Between the two sculptural segments, the great 355ft (108m)-long Beam of Light runs up the facade facing 42nd street and bursts out of the top of the building to pierce the sky. Closer to the ground, the shaft of the tower sits on a chunky base clad in a colorful abstract design. This portion of the building contains the 126 executive club rooms and suites before it drops down to street level, where there is a glass reception area.

The energy and vibrancy of the exterior is continued inside on a smaller scale. Here the distinctive color palette and geometric and abstract shapes seen on the outside are reworked and incorporated into textiles and wall murals.

There is bold abstract art, sleek black furnishings, contemporary furniture, and rooms are finished with luxurious materials. Along with guest rooms, the hotel has bars, a pastry shop, and a café, together with a spa and fitness center, more than 30 state-of-the-art meeting rooms, and no fewer than three ballrooms. Also part of the project at ground level is E Walk—a huge complex comprising stores, restaurants, a blues nightclub, and a movie theater.

Opened in October 2002, the Westin New York, which was designed by the architecture practice Arquitectonica, forms a key part of the regeneration of the previously run-down Times Square neighborhood. Along with theatergoers, the hotel serves a large business clientele which is drawn to the several million square feet of new office space that has been added to this area in recent years.

ARQUITECTONICA

Founded as an experimental studio in Miami in 1977, the architecture practice Arquitectonica has achieved renown worldwide for its exuberant, expressive, and colorful projects. Led by Bernardo Fort-Brescia and Laurinda Spear, its eyecatching schemes around the world have included glamorous apartment blocks, such as the much-loved Atlantis in Miami, with its swimming pool set in a frame midway up the building. They have also designed Miami City Ballet with its curvy facade, and the Embassy of the United States in Peru.

POWERING THE BEAM

Lighting the Beam of Light was a considerable technical accomplishment involving lighting technologies used in landscaping, sport, concert, and theater design. Incorporated into the hotel's facade are 466 10,000-hour-life lamps plus 466 Nite Star lighting fixtures (including 96 installed inside the hotel's atrium). The total wattage of the Beam of Light is 30,300 watts, equivalent to about 228 36in (91cm) television screens.

The dome is as high as a 27-story building.

The stadium seats 71,228 and cost $214 million to complete.

About 8,300 tons (7,530 tonnes) of reinforced steel, more than the weight of the iron and steel used in the Eiffel Tower in Paris, were used to build the Dome.

Georgia Dome

n March 1992, as workers fixed the last fabric panel in place, the Georgia Dome in downtown Atlanta laid claim to having the largest cable-supported fabric roof in the world. For seven years, until the Millennium Dome was constructed in London in 1999, it was the largest domed structure in the world.

The vast, multipurpose stadium is home to the city's football team, the Atlanta Falcons, but is also used for many other functions. It hosted the Super Bowl in 1994 and 2000, and during the 1996 Olympic Games was home to the basketball and gymnastics events. It is also used for trade shows, concerts, and political conventions.

Rock bands such as the Rolling Stones have played here to packed houses, and ministers, including Billy Graham and T. D. Jakes, have preached here.

The stadium was designed by architect Scot Bradley and structural engineer Matthys Levy. At ground level the building is rectangular, with chamfered glass corners, but as it rises a great hoop of concrete is raised on 52 column supports, and from here the fabric roof grows towards its highest point, supported on a complex network of steel

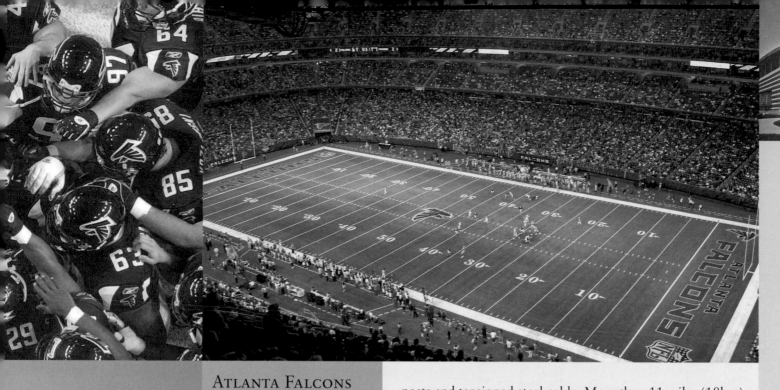

ATLANTA FALCONS

The Atlanta Falcons were founded in 1966, and the team name was chosen for the reasons given in a letter from schoolteacher Julia Elliott of Griffin, Georgia, who entered the contest to name the new team. She said that the name 'falcon' was appropriate as falcons are dignified and proud, with courage and fight.

posts and tensioned steel cable. More than 11 miles (18km) of supporting cable hold the roof, comprised of 130 Teflon-coated fiberglass panels, in place. The triangular and diamond-shaped panels, some as large as 80ft by 180ft (24m by 55m), were craned up through the web of cables and attached with aluminum fixings. To waterproof the roof, strips of fabric were then heat-welded over the seams.

A Clear View

One of the chief advantages of this innovative roof structure is that there are no columns or other supports to obscure the view of spectators. Spectators are also guaranteed a good view as seating is arranged in three cantilevered tiers and it is claimed that there isn't a bad seat in the whole stadium.

ROOF TECHNOLOGY

The roof design uses tensegrity, (from the two words "tensional" and "integrity"), a word coined by the world-renowned American engineer Richard Buckminster Fuller (1895–1983), who invented the geodesic dome in the 1940s. Tensegrity is an architectural system where structures stabilize themselves by balancing the counteracting forces of compression and tension. It gives shape and strength to both natural and artificial forms.

- The pylons of the bridge stand 242ft (74m) above the road deck and each supports a fan of 21 cables.
- The fabrication yard was 1.5 miles (2.5km) long and was the largest of its kind in the United States.
- A young woman with dripping wet clothes and hair is reputed to haunt the bridge.

Sunshine Skyway Bridge

Stretching for 5 miles (8km) across the mouth of Tampa Bay, the Sunshine Skyway is one of the longest bridges in the world. Some of this enormous length is accounted for by undistinguished approaches over land or shallow water, but the 4-mile (6km) central section is quite another matter, with a 22-mile (3km) stretch of high-level spans over the main navigation channel to the port of Tampa.

Built to replace an older bridge damaged in a catastrophic accident in 1980, the Sunshine Skyway has tremendous verve and style, and was daringly original at the time of its completion in 1987. Designed by Figg Engineering, it was one of the first bridges ever to employ single fans of cables fastened to the center of the road deck for support. This elegant design is now a well-established method of construction, but at the time it was virtually untried. In recognition of the image of "the Sunshine State," the stays are painted brilliant yellow, and at night they are illuminated with a golden glow to accentuate their form. Linking major holiday resorts along Florida's Gulf coast and spanning busy shipping lanes, the bridge looks equally impressive from both the highway and the water.

Assembling the Parts

A cable-stayed bridge can never match the massive spans achieved by traditional suspension bridges, but it is far simpler to construct. The deck, which forms part of the load-bearing structure, is usually built up from prefabricated segments, which are fastened to the cables for additional support. The Sunshine Skyway has a hollow concrete deck, 95ft (29m) wide and 14ft (4m) deep, that is made up from segments weighing up to 220 tons (200 tonnes) apiece; the largest that had ever been erected on a segmental concrete bridge. The last piece was lifted into place in August 1986, completing the 1,200ft (366m) main span, which forms a graceful sweep 190ft (58m) above the sea. Completed at a cost of $245 million, the bridge has won many awards for its design.

There have, however, been some problems with the materials, such as corrosion of the steel reinforcing bars in some of its supporting piers. Innovation, unfortunately, is never without risk, but although repairs may well prove costly, the Sunshine Skyway still deserves to be regarded as a masterpiece of engineering style.

SURVIVING THE STORMS

In 2005 Florida was buffeted by more than 20 tropical storms and hurricanes, making it the worst year for storms since records began in 1851. Exceptionally high winds have swept the Sunshine Skyway on several occasions, forcing temporary closure to all traffic. So far, however, no damage has been caused, since the bridge was designed to cope with such extreme conditions.

Long before construction work started, specialist consultants built a 1:80 model of the deck and a 1:375 model of the entire bridge. These were tested in a wind tunnel to gauge the effects of both turbulence and pressure. On both accounts the bridge remained unaffected in winds of well above 100mph (160kph).

COLLISION COURSE

The present bridge is the second on the site. At 7.38am on 9 May, 1980 the freighter *Summit Venture* collided with the older structure during a sudden violent squall, causing 1,000ft (300m) of the roadway to collapse. Eight cars and a Greyhound bus fell 150ft (46m) into the sea, causing the loss of 35 lives. On the new bridge, the six central piers are protected by artificial islands, known as dolphins, which are designed to withstand the impact of an 87,000-ton ship traveling at 10 knots.

the facts

- The building's internal volume is equivalent to 3.5 Empire State Buildings.

- In 2004, Hurricane Frances tore 1,000 aluminum panels from the building, leaving the interior exposed.

- Launch Complexes 40 and 41, on an adjoining US Air Force base, are used for unmanned space flights.

97 Kennedy Space Center   **Cape Canaveral**, USA

Kennedy Space Center

As the main base for America's space program, the Kennedy Space Center at Cape Canaveral has witnessed the greatest technological adventure of the modern age. Covering 140,000 acres (57,000ha) of low-lying sand and silt on Florida's Atlantic coast, the NASA base is a vast, empty wasteland, with most activities concentrated in an area of a few square miles known as Launch Complex 39.

Built between 1963 and 1965 at a cost of $800 million, this is the world's foremost space-port, employing 17,000 workers, and containing an extraordinary array of highly specialized facilities that includes one of the largest buildings in the world. Originally designed to house up to four of the 363ft (111m) Saturn V rockets that took the manned Apollo missions to the Moon, the Vehicle Assembly Building is constructed on a truly mind-boggling scale. Standing 525ft (160m) high, the building measures 715ft (218m) long by 518ft (158m) wide, providing a floor-space of 8 acres (3ha). Its enormous doors, which consist of 11 sections that take almost an hour to open, stand 456ft (139m) high, three times as tall as the Statue of Liberty. The curious shape of the doors, with extra-wide sections at ground level, is designed to admit the giant transporters that transfer space craft from the building to their launch-pads.

Assembling the Shuttle

Since 1981 the building has been used to assemble space shuttles before they are launched. The interior is divided into four high bays, each the full height of the building, and a low bay aisle for general maintenance. Many months before the designated launch date, one of the three remaining orbiters, *Discovery*, *Atlantis* or *Endeavour*, is

CRAWLER TRANSPORTERS

The assembled shuttle, resting on its mobile launch platform, is carried to the launch pad on board one of NASA's two Crawler Transporters, the largest such vehicles in the world. Measuring 130ft (40m) by 115ft (35m), each transporter weighs 2,998 tons (2,720 tonnes) and moves on four units of twin caterpillar tracks standing almost 10ft (3m) high. Its two 2,750 horsepower diesel generators power 16 electric motors, consuming 150 gallons (568 liters) of fuel per mile. With the shuttle and its launch platform weighing 6,834 tons (6,200 tonnes), the 4-mile (6km) journey to the launch pad can take well over five hours.

towed over from the nearby hangers of the Orbiter Processing Facility and raised to a vertical position. Meanwhile, in other bays, the craft's huge external fuel tank is checked over and its two Solid Rocket Boosters are assembled on a mobile launch platform. Finally, the shuttle's separate elements are put together, with assistance from the building's 73 overhead cranes, and checked in readiness for transportation to the launch pad.

Launches are spectacular events that draw enormous crowds and, while access to much of the Space Center is restricted, the extensive visitor complex attracts more than 2 million people a year. Exhibits that include enormous rockets alongside spacecraft from the Mercury, Gemini, Apollo, and Shuttle programs bear witness both to NASA's engineering triumphs and to the bravery of astronauts who have taken the first steps of man's journey into space.

SHUTTLE PROGRAM

The Shuttle Program was initiated in the 1970s to provide a low-cost, reusable spacecraft that could carry heavy loads to space stations orbiting earth. The first operational flight was in 1981, but the program has since suffered two catastrophes with the destruction of *Challenger* in 1980 and *Columbia* in 2003. Despite some problems with *Discovery*'s mission in 2005, the program is scheduled to continue until 2010, playing an essential role in the completion of the United States-Russian International Space Station. Although the space station has been permanently manned since November 2000, it is still less than half the intended size.

Designing and building the platform took 15 million man hours.

Thunder Horse displaces 143,000 tons (130,000 tonnes) of seawater.

The platform's impact on the environment is lowered as it recovers waste heat, reducing the energy bill.

Thunder Horse Oil Platform

Despite an almost universal recognition of the need to find alternatives to fossil fuels, the world's insatiable demand for oil and gas continues to gain pace, driving oil companies to search for fresh reserves under ever more challenging conditions. The Thunder Horse Platform in the Gulf of Mexico is an astonishing example of how this challenge has stimulated engineers to build increasingly massive and sophisticated structures.

The oil field, discovered 150 miles (240km) south of the Louisiana coast by BP in 1999, is the largest ever found in the region, but tapping into this rich reservoir poses problems that can scarcely be imagined. The oil lies buried under 20,000ft (6,000m) of rock and mud beneath a sea 6,000ft (1,830m) deep. When drilled from such a depth, oil emerges at a pressure of 17,400 psi and at a temperature of 275°F (135°C)–conditions never previously encountered on any offshore rig. It was these extreme demands that led to the construction of Thunder Horse, the largest and most innovative floating oil platform in the world.

With a deck the size of three soccer pitches, the platform is essentially a giant raft that floats on submerged watertight pontoons. The main structure, supported on four corner columns, is a rectangular steel box measuring 446ft (136m) long, 365ft (111m) wide, and 33ft (10m) high. It contains living quarters for a workforce of 139, while operational equipment is housed in modules alongside the helipad and drilling derrick on the open deck.

Linked into a network of submarine pipelines, Thunder Horse is designed to produce up to 250,000 barrels of oil and 7 million cubic yards (5.6 million cu m) of natural gas a day, but, unlike most offshore rigs, it incorporates a range of innovative features designed to minimize environmental harm. Sand is brought up to the surface by pumps, then shipped to shore for cleaning and recycling, while waste water, instead of being flushed into the ocean, is mixed with seawater and then injected back into the oilfield to maintain the reservoir's high pressure.

DIFFERENT PLATFORMS

Offshore oil platforms fall into two categories: development platforms (for exploration and drilling work, but not processing), and production platforms (with all the processing plant required to keep wells in production for a long period). Thunder Horse is a production platform. Both types of platform come in a variety of structural designs. In shallower waters, jack-up rigs have legs extending down to the seabed. Semisubmersible rigs, such as Thunder Horse, float and maintain their stability with underwater pontoons. The Horse is anchored in position by 16 chain-, wire- and suction-pile anchors.

BLACK GOLD

BP discovered this oilfield by sending the drillship *Discoverer 534* to sink a "discovery well" in 1999. Enough oil was found to have the drillship *Discoverer Enterprise* send down an appraisal well. *Discoverer 534* then found more oil farther north, so BP starting building the production platform. Despite risks that the oil estimates are wrong or that oil prices will collapse, the potential profit is enormous. Thunder Horse is expected to handle at least 1 billion barrels of oil over the next 25 years and, after Hurricane Katrina struck in 2005, prices rocketed, making Thunder Horse's "black gold" worth $70 billion on that day.

- The building contains 88,504sq yards (74,000sq m) of office space, with an observation deck on floor 52.
- The largest dampers measure 6ft by 2ft (2m by 1m) and have a thrust of 1,260,000lbs (570,000kg).
- Diamond-shaped trusses spread stresses more evenly than regular X-shapes, enhancing structural stability.

Torre Mayor

Towering above its neighbors on Mexico City's fashionable Paseo de la Reforma, the Torre Mayor is a thrilling sight. The building is a daring mix of styles, with a curving cliff of glass rising up its main facade and a tower of polished stone behind. But despite the subtle details, the complex roofline, and the sculpted forms, the real excitement comes from the tower's enormous height.

At 738ft (225m), it became Latin America's tallest building when it was completed in 2003, far outstripping any rival in the low-rise sprawl of Mexico's enormous capital city. In fact, prior to its construction, no building of more than 38 stories was permitted in Mexico City, which lies in a highly active earthquake zone and imposes stringent seismic regulations on all new developments. In this respect at least, the 55-story Torre Mayor is not quite as risky it might appear. It did not break the rules, but simply pushed them to new limits through the application of brilliant engineering. Using systems that were first developed to protect US missile silos from nuclear attack, it is designed not merely to withstand an earthquake with a magnitude of 8.5, but to remain virtually unscathed. Long before construction started in 1998, geo-technical surveys were undertaken to analyze the site's soft, unstable soils, and 3-D computer models were created to test methods of absorbing seismic shocks. The tower's first line of defense lies in its foundations, which rest on a concrete mat up to 8ft (3m) thick.

Built to Last

The structure of the tower is essentially a steel frame, with a perimeter of columns and diamond-shaped trusses that are tied through the floor-spans to a central spine. Up to the 30th floor, the columns are encased in concrete for extra strength, while above this, the lighter and more flexible steel is left exposed. But the most amazing innovation lies in the "viscous fluid damping system." Ninety-eight giant pistons, like the shock absorbers on a car but thousands of times more powerful, are built into the spine and frame.

Self Sufficient

The tower is a remarkably self-contained environment with its own shopping center, restaurants, integrated parking and rooftop helipad. The 1in (2.5cm)-thick glass skin is hermetically sealed to exclude dust and pollution, with the air inside filtered six times every hour to remove suspended particles. What is more, all of the building's mechanical and electronic systems, including lighting, air-conditioning, elevators, fire alarms, and even washroom facilities, are integrated by the tower's sophisticated "brain," a computerized Building Management System (BSM). The BSM is programmed to respond to almost any conceivable event, from an earthquake to a dripping faucet.

DANGER ZONE

Mexico City is built on an active fault-line between tectonic plates, and its soft clay soil tends to amplify seismic waves. In September 1985 a catastrophic earthquake with a magnitude of 8.1 destroyed 100,000 homes and left up to 20,000 dead in just three minutes. More recently a quake with a magnitude of 7 struck in June 1999, and in January 2003 a shock measuring 7.6 shook the city for some 30 seconds, causing widespread damage. In the newly finished Torre Mayor, office workers saw the shock-absorbers start to pump and light fittings swayed, but the building, as expected, was unharmed.

The reservoir covers 525sq miles (1,350sq km).

The dam contains 15 million cubic yards (12 million cu m) of concrete, 15 times the amount used in the Channel Tunnel between England and France.

Water flows through the turbines at the rate of 167 million gallons (760 million liters) an hour.

Itaipú Dam

With global electricity consumption increasing every year, hydro-electric schemes have an important role to play in the search for clean and sustainable sources of power. The awesome force of the Paraná River, on the border between Brazil and Paraguay, is spectacularly evident at the Iguassu Falls, the mightiest waterfalls in South America. Just a few miles upstream this same force has been harnessed by the almost equally spectacular Itaipú Dam, which remains the world's

A Joint Project

The dam is owned equally by Brazil and Paraguay, with a line through the control room marking the national frontier. Since Brazil's energy requirements far exceed those of its much smaller neighbor, Paraguay sells on most of the electricity generated by its turbines on the right bank of the river, making it the world's largest exporter of hydro-electric power. Two overhead lines carrying 600 kV each run from Foz do Iguacu in Paraguay, 500 miles (800km) to Sao Paulo in Brazil, but since the countries use different systems, the power has to be converted from 50Hz to 60Hz.

PARANÁ RIVER

With a length of 1,600 miles (2,580km), the Paraná is the second-longest river in South America, draining an area of almost a million square miles (2.5 million sq km) in Brazil, Paraguay and Argentina. In 1998 Paraguay and Argentina completed the construction of a dam at Yacireta, near the city of Encarnacion. Although far smaller than Itaipú, the project affects a densely populated region and has proved highly controversial. Some 40,000 people have been relocated and 80,000 more may have to move if plans to raise the water level by a further 23ft (7m) are put into effect.

largest hydro-electric scheme until China's Three Gorges project is completed. Generating 700 megawatts apiece, each of Itaipú's 18 massive turbines equals the output of a medium-sized nuclear reactor. At full capacity, the scheme can produce up to 75 trillion watt-hours of electricity in a year, supplying 15 percent of Brazil's requirements and almost all those of Paraguay. If this same amount of energy was generated by coal-fuelled power stations, their carbon dioxide emissions would amount to 67.5 million tons (61 million tonnes). The dam has also proved to be a popular attraction with half a million visitors a year. It is floodlit every night, creating an awe-inspiring scene as spray rises from the water that thunders down the slipways.

A Line of Dams

The project was a massively ambitious undertaking that took 16 years to finish, cost $20 billion, and involved up to 30,000 workers on site at the peak period of construction. When work began in 1975, the first three years were spent digging a new course for the Paraná River to divert its flow around the site. This involved shifting 50 million tons (45 million tonnes) of rock and earth to create a channel 1 mile (2km) long. Four linked but separate dams were then built across the river valley, stretching for a total length of 5 miles (8km). Each one is constructed differently, with some filled with earth or stone, while the main dam is a hollow gravity dam made from reinforced concrete.

The first turbine came on line in December 1983 and by March 1991, 18 were in operation, with a further two planned for completion in 2005, bringing the scheme's total output to 14,000 megawatts. Behind the dam a huge reservoir, 110 miles (170km) long, has submerged what was previously forest. In its place 20 million saplings have been planted and a wildlife reserve established.

Index

Glossary

abutment lateral bridge support

apex top, tip

asymmetric without symmetry

atrium central courtyard

box girder hollow main support made of plates bolted together

brace strengthening support within a structure

cable stay bridge the roadway is held up by tensioned wire cables suspended from the top of a tower

caisson chamber for laying underwater foundations, open underneath so that air pressure keeps the water out

cantilever bridge the roadway is supported on long girders projecting from either side of a pier or pier

catenary curve the curve formed when a chain is suspended from two points or towers of unequal height

cladding external covering on a building

conduit channel or pipe for the protection of cables

derrick type of crane used for shifting heavy loads

exoskeleton external main support frame

geodesic dome stable dome-shaped building or structure, using short struts and small triangular planes, created along mathematical principles

G-force acceleration due to gravity

gravity dam the strength of the dam is due to the sheer weight of the materials from which it is built

hydrostatic tension – tension reliant on the balance of liquids

mullion vertical bar dividing panes in a window

nacelle outer casing of an aircraft engine

parabola curve formed by the intersection of a cone with a plane parallel to its side

parterre level space in a garden

pile heavy beam driven straight into the ground as an architectural support

pre-stressed concrete concrete strengthened by the addition of wires

sluice moveable gate to control the flow of water

suspension bridge the roadway is suspended from cables that pass over towers and are anchored in the ground

truss supporting structure

tuned mass damper sophisticated shock absorber

vortex whirling particles or mass

Acknowledgements

Abbreviations for terms appearing below: (t) top; (b) bottom; (c) center; (l) left; (r) right.
The Automobile Association wishes to thank the following photographers, companies, and picture libraries for their assistance in the preparation of this book.

© 2001 Karant + Associates, Inc. 188cl; **Alamy.com** 4bl (B&Y Photography), 19tr (Arcblue), 19br (The Hoberman Collection), 49b (F1 online); 59br (mediacolor's), 61 (Jon Arnold Images), 61tr (Jon Arnold Images), 76/7 (Arcaid), 89br (F1 online), 94l (Panorama Media (Beijing) Ltd), 97tr (Panorama Media (Beijing) Ltd), 99cr (Iain Masterton), 109 (Rollie Rodriguez), 118/9 (Philip Corbluth), 119tr (chromepix.com), 120r (Nic Cleave Photography), 120/1 (Bill Rubie), 121l (Nic Cleave Photography), 121tr (Nic Cleave Photography), 122/3 (Aflo Foto Agency), 123tr (Photo Japan), 123br (Photo Japan), 127tc (Cris Haigh), 127tr (J Marshall – Tribaleye Images), 130bc (B&Y Photography, 136/7 (Glen Allison), 144/5 (Gavin Hellier), 148c (Jon Arnold Images), 149tr (Worldwide Picture Library), 149tc (Jon Arnold Images), 170 (Winston Fraser), 190/1 (Andre Jenny), 191tc (Andre Jenny), 191tr (Andre Jenny), 213tc (David R Frazier Photolibrary, Inc.), 212/3 (Andre Jenny); **Alnwick Garden** 16/17, 17tl, 17tr, 17c, 17br; **American Museum of Natural History** 206 (Denis Finnin), 206/207 (Denis Finnin), 207tl (Roderick Mickens), 207tcl, 207tr (Denis Finnin), 207tcr (Roderick Mickens); **Sarah Anderson/Keck Observatory** 164tr, 172, 173tr; **Paul Andreu Architecte** 85tc, 85tr, 85cr; **Arcaid** 98 (C Y Lee and Associates/Marc Gerritsen), 99 (C Y Lee and Associates/Marc Gerritsen), 99tr (C Y Lee and Associates/Marc Gerritsen), 107 (Kenzo Tange/Bill Tingey), 107tr (Kenzo Tange/Bill Tingey); **Arken Museum** 6tr, 39tc, 39tr; **Andy Arthur Photography** 60/1; **Atlanta Falcons** 211tl, 211tc; **Auspic, supplied courtesy of Department of Parliamentary Services** 136, 137tr, 137cr; **Axiom Photographic Agency** 97br; **Bill Bachman** 140c; **Baiyoke Sky Hotel Bangkok Thailand, www.baiyokehotel.com** 126; **BP** 216, 217, 217tr, 217br; **British Airways London Eye** 6tcl, 20/21; **British Waterways Photo Library** 5tc, 11tr; **Cedar Point** 165tr, 192bl, 192tr, 193, 193tr; **Central Japan Railways** 112; © Copyright **CERN** 5bc, 7tcl, 56, 56/57, 57tr, 57cr, 57b; **Copyright Commonwealth of Australia. Reproduced with permission from Snowy Hydro Limited under licence from National Archives of Australia** 138 tr; **Corbis UK** 148/9 (Lloyd Cluff), 187 (Alan Schein Photography), 187cr (Alan Schein Photography), 199cr (James Leynse), 200 (Alan Schein Photography), 201tr (Alan Schein Photography); **CPA Media/David Henley** 88/9, 89tr, 129tl, 129tcl; **Hayes Davidson and John Maclean** 26/27b, 26/27t, 27tl; **Deutsche Bahn AG** 42/43 (Jazbec), 42 (Lautenschläger), 43tr (Lautenschläger); **Eden Project** 6tcr (Apex/Simon Burt), 29tl, 29tr; **Emaar** 143tcl, 158/9, 159tl, 159bl, 159tr; © **Empics** 18/9t, 93tcl, 94r, 203cr, **Energi E2** 36/37, 37cl, 37c, 37tc, 37tr; **Esto** 102/3 (© Tim Griffith), 103 (© Tim Griffith), 201tl (© Scott Frances), 201tcl (Wolfgang Hoyt), 201tcr (© Scott Frances); **Figg, Engineer of Record** 212, 213tl, 213tr; **Erika Barahona-Ede ©FMGB Guggenheim Bilbao Museoa, 2006, All Rights Reserved** 72, 73tr; **Foster and Partners** endpapers, 47br, 80tcr, 82t, 82cr, 83, 83tr; **Gateshead Council** 7tr, 12, 12/13, 13tr, 14/15; **Gehry Partners, LLP** 179cl; **Georgia Dome** 164tl, 211tr, 211br; **Getty Images** 18/9b, 19tc (Photographers Choice), 32, 32/3, 33tc, 33tr, 55tc (The Image Bank), 64, 64/5, 65tc, 65tr, 66/7, 88 (Taxi), 90/1, 100 (Taxi), 101 (Taxi), 101tr (Taxi), 104 (Taxi), 104/5 (Panoramic Images), 106, 111 (AFP), 112/3, 113tl (Taxi), 113tr (Taxi), 122 (AFP), 130/1 (Taxi), 130cl (Image Bank), 144 (AFP), 147tc, 156br (Robert Harding Picture Library), 159cl, 161br (AFP), 162/3 (Tim Graham), 171tc, 184/5, 185cr (Image Bank), 202, 202/3, 203tr, 221tl (AFP); **Alain Goustard** 62/3, 63tl, 63tc, 63tr; **Doug Hall of i2i, Images courtesy of Gateshead Council** 14, 15tc, 15tr; **Haystac** 140r, 141, 141cr, 141tr; **Hijjas Kasturi Associates** 81tcr, 132, 132/133, 133 tr, 133 cr; **Hollandse Hoogte** 40/1, 41tr; **Hotel Arts Barcelona** 75tl, 75tc; **afd. Imaginechina** 86/7, 87tr, 91cr, 93tl, 93tcr, 93cr, 96/7, © **Jewish Museum Berlin** 48/49 (Jens Ziehe, Berlin), 48 Menashe Kadishman Shalechet (Fallen Leaves) 1997-2001. Long-term loan from Dieter and Si Rosenkranz, Berlin (Photo; Jens Ziehe, Berlin) 49tr (Jens Ziehe, Berlin); **Jumeriah International** 4tc, 5tr, 143tcr, 143tr, 152, 153, 153tl, 154/155, 155tl, 155cl, 155tr, 155cr, 155br; **Kansai International Airport** 80tcl, 116; **W M Keck Observatory** 173; **Keystone** 59tr; **Kingdom Center** 3, 142tcr, 150, 150/1c, 150/1, 151tc, 151tr; **Yves Klein Fuente de fuego (Fire Fountain), 1961 Photo: Erika Barahona-Ede ©FMGBGuggenheim Bilbao Museoa, 2005** 4btl; **Ian Lambot** 55, 55tr; **Hervé Langlais/Paul Andreu Architecte** 5bl, 80tl, 84/5; **Luftbild Schweiz** 58/9; **Magnum Photos** 60l (© Jean Gaumy); **Massachusetts Turnpike Authority** 5br, 165tcr, 196/7, 197tl, 197tr; **Anthony May Photography** 188c, 189, 189tr; **Satoru Mishima/Foreign Office Architects** 114/115, 115tl, 115cl, 115tr; **Nakheel** 143tl, 157tr, 157, 160/161, 161tr, 161tl, 161tc; **NASA** 214, 214/5, 215tr, 215c; **Isaac Newton Group, La Palma** 78/79 (Nik Szymanek), 79tc, 79tr, 79cl; **Norconsult** 129cl, 129cr; **Øresundsbro Konsortiet** 35tc, 35c, 35bc; **Keith Paisley, images courtesy of Gateshead Council** 13tc, 13tcr; **Pan Pacific Vancouver** 167cr; **Graeme Peacock** 15c; **Photo Japan** 120l (Kenneth Hamm); **Photolibrary** 157tl (Joe Malone); **Pictures Colour Library** 23br, 28/9, 70/1t, 70/1b, 71tr, 76, 184, **Plan Architect** 127l; **QA Photos Ltd** 30, 30/1, 31tr, 31cr; **Reichmann International, Mexico** 218bc, 218cr, 219, 219tr, 219br; **Renewable Energy Systems (RES)** 182 (Hendershot Photography), 182/183 (Hendershot Photography), 183bl, 183tc (Hendershot Photography), 183tr (Hendershot Photography); **Reuters** 86 (Claro Cortes IV), 220/1 (STR), 221tc (Paulo Whitaker), 221tr (STR); **Rex Features Ltd** 27bc, 95, 97cr, 111br, 156bc; **Rock and Roll Hall of Fame and Museum** 190; C1995, C1995 Kevin C. Rose/AtlantaPhotos.com 210/1; **RWS MD**; **afd. Multimedia/Waterland Neeltje Jans** 41tc; © Scottish Parliamentary Corporate Body 2005 5tl, 6tl, 8/9, 9tr, 9c, 9cr; **Sinopix** 102, 103tr; **Siripong Kanjanabut/OnAsia.com** 128/9; **Skidmore, Owings & Merrill LLP, Architects** 5cr (dbox), 159, 204 (dbox), 205 tl (dbox), 205tr (dbox), 205c (dbox); Copyright **Snowy Hydro Limited 2005** 80tr, 138cl, 139tl, 139tr, 139; **Sony Berlin GmbH** 50; **David S Steele** 180/181; **Lori Stiles/University of Arizona** 181bl, 181tc, 181tr; **The Stubbins Associates, Inc** 198, 199, 199tr; **Sydney Opera House** 81tl, 135tl, 135, 135tr; **Tokyo International Forum** 4br, 81tcl, 108, 109tr; **Travel Ink/Chris Stock** 125br; **Tunnel du Mont Blanc** 68, 68/69, 69tr, 69cr, 69bc; **Vancouver Convention & Exhibition Centre** 166br; **Sandra Vannini** 146/147; **View Pictures Ltd** 10/1 (Hufton & Crow), 11br (Hufton & Crow), 25 (Grant Smith), 38/9 (Dirk Robber/Artur), 39tl (Dirk Robber/Artur), 52/3 (Dennis Gilbert), 53tl (Dennis Gilbert), 53tr (Dennis Gilbert), 53br (Dennis Gilbert), 54/5 (Dennis Gilbert), 105 (Dennis Gilbert), 105 tr (Dennis Gilbert), 116/7 (Dennis Gilbert), 117 (Dennis Gilbert), 117tr (Dennis Gilbert), 178 (Hufton & Crow), 179cr (Hufton & Crow), 186 (Nick Meers), 187tr (Nick Meers); **Walt Disney Concert Hall at the Music Center of Los Angeles County** 4tr (Courtesy of the Music Center of Los Angeles County), 164tcr (Courtesy of the Music Center of Los Angeles County), 176cr (Howard Pasamanick), 176cb (Federico Zignani), 176/177 (Courtesy of the Music Center of Los Angeles County), 179tl (Tim Street-Porter), 179tr (Courtesy of the Music Center of Los Angeles County); **Wembley National Stadium Ltd** 22, 22/23, 23tl, 23tr; **Westin Hotel** 208, 209, 209tr; **West Virginia Tourism** 165tcl, 194/5, 195tr, 195bl; **Nick Wood Photographer** 20bl, 20bcl, 20bcr, 20br; **World illustrated/photoshot** 34/35, 40, 91tr, 92/93, 93tr; **World Pictures** 110, 111tr, 118, 142tcl, 156cl, 179tr, 185tr; **Nigel Young/Foster and Partners** 25tl, 25tr, 45cr, 47l; **YRM Architects** 142tr, 162, 163tl, 163tr; **Yukio Yoshimura** 81tr, 124,124/125, 125tr; **Gerald Zugmann/SNØHETTA** 147tl, 147tr.

The remaining photographs are held in the Association's own photo library (AA World Travel Library) and were taken by the following photographers: **Pete Bennett** 213tl; **Ian Burgum** 142tl, 145tl, 145tr; **Michelle Chaplow** 75tr; **Steve Day** 74/5; **Max Jourdan** 7tcr, 21tr, 66, 72/3, 165tl, 174, 174/5, 175tr, 175br; **Mike Langford** 134c; **Simon McBride** 7tl, 44, 44/5, 46/7, 47tr, 51, 51tr, 51bl; **Anna Mockford & Nick Bonetti** 4bc, 77tc, 77tr; **Jean François Pins** 169, 169tr, 171b; **Clive Sawyer** 24/5, 164tcl, 166cl, 166/7, 167tr; **Neil Setchfield** 131tr; **Tony Souter** 65cr; **Nick Sumner** 168cl, 168bl, 168cr, 171tl, 171tr; **Wyn Voysey** 67tr; **Phil Wood** 189cr.

Every effort has been made to trace the copyright holders, and we apologise in advance for any accidental errors. We would be happy to apply the corrections in the following edition of this publication.